Advances in Ceramic Matrix Composites VIII

Related titles published by The American Ceramic Society:

Innovative Processing and Synthesis of Ceramics, Glasses, and Composites VI (Ceramic Transactions Volume 135)
Edited by Narottam P. Bansal and J.P. Singh
©2002, ISBN 1-57498-150-1

Innovative Processing and Synthesis of Ceramics, Glasses, and Composites V (Ceramic Transactions Volume 129)
Edited by Narottam P. Bansal and J.P. Singh
©2002, ISBN 1-57498-137-4

Advances in Ceramic Matrix Composites VII (Ceramic Transactions Volume 128)
Edited by Narottam P. Bansal, J.P. Singh, and H.-T. Lin
©2001, ISBN 1-57498-136-6

Advances in Ceramic Matrix Composites VI (Ceramic Transactions Volume 124)
Edited by J.P. Singh, Narottam P. Bansal, and Ersan Ustundag
©2001, ISBN 1-57498-123-4

Innovative Processing and Synthesis of Ceramics, Glasses, and Composites IV (Ceramic Transactions Volume 115)
Edited by Narottam P. Bansal and J.P. Singh
©2000, ISBN 1-57498-111-0

Innovative Processing and Synthesis of Ceramics, Glasses, and Composites III (Ceramic Transactions Volume 108)
Edited by J.P. Singh, Narottam P. Bansal, and Koichi Niihara
©2000, ISBN 1-57498-095-5

Advances in Ceramic Matrix Composites V (Ceramic Transactions Volume 103)
Edited by Narottam P. Bansal, J.P. Singh, and Ersan Ustundag
©2000, ISBN 1-57498-089-0

The Magic of Ceramics
By David W. Richerson
©2000, ISBN 1-57498-050-5

Advances in Ceramic Matrix Composites IV (Ceramic Transactions Volume 96)
Edited by J.P. Singh and Narottam P. Bansal
©1999, 1-57498-059-9

Innovative Processing and Synthesis of Ceramics, Glasses, and Composites II (Ceramic Transactions Volume 94)
Edited by Narottam P. Bansal and J.P. Singh
©1999, ISBN 1-57498-060-2

Innovative Processing and Synthesis of Ceramics, Glasses, and Composites (Ceramic Transactions Volume 85)
Edited by Narottam P. Bansal, Kathryn V. Logan, and J.P. Singh
©1998, ISBN 1-57498-030-0

For information on ordering titles published by The American Ceramic Society, or to request a publications catalog, please contact our Customer Service Department at 614-794-5890 (phone), 614-794-5892 (fax), <customersrvc@acers.org> (e-mail), or write to Customer Service Department, 735 Ceramic Place, Westerville, OH 43081, USA.

Visit our on-line book catalog at <www.ceramics.org>.

Advances in Ceramic Matrix Composites VIII

Ceramic Transactions
Volume 139

Proceedings of the Ceramic Matrix Composites Symposium at the 104th annual meeting of The American Ceramic Society, April 28–May 1, 2002, in St. Louis, Missouri.

Edited by
J.P. Singh
Argonne National Laboratory

Narottam P. Bansal
National Aeronautics and Space Administration
Glenn Research Center

M. Singh
QSS Group, Inc.
NASA Glenn Research Center

Published by
The American Ceramic Society
735 Ceramic Place
Westerville, Ohio 43081
www.ceramics.org

Proceedings of the Ceramic Matrix Composites Symposium at the 104th annual meeting of The American Ceramic Society, April 28–May 1, 2002, in St. Louis, Missouri.

Cover photo: "SEM photomicrograph of transverse cross section of typical $ZrSiO_4$-based FM" is courtesy of K.C. Goretta, F. Gutierrez-Mora, T. Tran, J. Katz, J.L. Routbort, T.S. Orlova, and A.R. de Arellano-López, and appears as figure 1 in their paper "Solid-Particle Erosion of $ZrSiO_4$ Fibrous Monoliths," which begins on page 139.

For information on ordering titles published by The American Ceramic Society, or to request a publications catalog, please call 614-794-5890.

Printed in the United States of America.

4 3 2 1–05 04 03 02

ISSN 1042-1122
ISBN 1-57498-154-4

Contents

Processing–Microstructure–Property Relationships

Mechanical Behavior

Characterization

Preface

Ceramic composites are leading candidate materials for high-temperature structural applications. This proceedings volume contains papers presented at the symposium on ceramic-matrix composites held during the 104th Annual Meeting and Exposition of The American Ceramic Society (ACerS), April 28–May 1, 2002, in St. Louis, Missouri. This symposium provided an international forum for scientists and engineers to discuss all aspects of ceramic composites. A total of 48 papers, including invited talks, oral presentations, and posters, were presented from 14 countries (the United States, Australia, Belarus, Belgium, Brazil, Canada, Germany, Japan, Kyrgzstan, the People's Republic of China, Russian Federation, Spain, Taiwan, and Turkey). The speakers represented universities, industry, and government research laboratories.

This volume contains thirteen invited and contributed papers, all peer-reviewed according to ACerS procedures, on the latest developments in processing and fabrication methods, process modeling, processing–microstructure–property relationships, mechanical behavior, and characterization. The result is that all of the most important aspects necessary for the understanding and further development of ceramic composites are discussed.

The organizers are grateful to all participants and session chairs for their time and effort, to authors for their timely submissions and revisions of the manuscripts, and to reviewers for their valuable comments and suggestions; without the contributions of all involved, this volume would not have been possible. Financial support from ACerS is gratefully acknowledged. Thanks are due to the staff of the Meetings and Publications Departments of ACerS for their tireless efforts. We especially appreciate the helpful assistance and cooperation of Greg Geiger and Sarah Godby throughout the production process of this volume.

J.P. Singh

Narottam P. Bansal

M. Singh

rocessing and Fabrication Methods

CERAMIC MATRIX– AND LAYER–COMPOSITES IN ADVANCED AUTOMOBILE TECHNOLOGY

Michael Buchmann, Rainer Gadow*, Dietmar Scherer and Marcus Speicher
Institute for Manufacturing Technologies of Ceramic Components and Composites, IMTCCC
University of Stuttgart
Allmandring 7b
D-70569 Stuttgart
GERMANY

ABSTRACT

Bulk ceramic materials for structural and functional applications have been in the center of interest in R&D and product development of various industries since many years. Even if the technical requirements could be achieved the cost/performance ratio was unsatisfying in many cases. Recent product development in light weight design and engineering offers a lot of interesting manufacturing and operation features if cost effective composite solutions including ceramic components can be realized. Advanced light metal engineering and ceramics must not be competitors or contrary but in an appropriate combination they can excel in production cost, functionality, performance and energy consumption. Ceramic matrix composites with carbon fiber reinforcement can fulfill these requirements for high temperature, structural and friction applications and the same is true if light metal substrates can be combined with performing ceramic, cermet, metallurgical or polymer based surface layers for various machine elements and system components. Protective and functional coatings on light metal surfaces can be individually designed.

The paper describes manufacturing technologies as well as materials and components characterization results, e. g. high temperature resistant friction materials made from SiC based CMC as well as coating technologies for the deposition of ceramic, metallurgical and preceramic polymer coatings on various light metal substrates. The specific characteristics and advantages of these composites are presented for advanced disk brake technology and full light metal motor engineering with a focus on ceramic coated crankcases and cylinder surfaces of combustion engines. In addition, tribologically optimized combined coatings on light metal components are introduced. Their tailored surface properties can compete with bulk ceramics and feature with additional characteristics like antiadhesive and dry lubricant ability.

INTRODUCTION

In the last decade there was an increasing effort on weight reduction in the automotive industry, e.g. in ultra lightweight structures for high performance sports cars[1]. Concerning the applied materials one main focus has been on polymer matrix composites (PMC) like carbon and glass fiber reinforced plastics (CFRP, GFRP resp.) and light metals like aluminum and magnesium (Fig. 1).

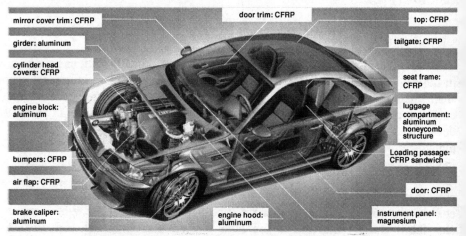

Fig. 1 Lightweight construction on the basis of serial BMW M3 [2]

Further potential for weight savings can be provided by ceramic matrix composites (CMC). The substitution of cast iron brake disks by carbon fiber reinforced silicon carbide ceramics leads to mass reduction in the range of 30 kg per vehicle. The excellent tribological properties and thermal stability of these CMCs offer the possibility for a life time application and improved braking performance. But for a series application of these ceramic brake disks the processing times, i.e. manufacturing costs, have to be lowered.

The use of light metals is mostly limited by their poor oxidation and wear resistance thus resulting in the strong demand for an additional protective or functional surface treatment. Thermally sprayed ceramic, hard metal and metallurgical coatings can optimize the wear resistance even under harsh conditions in the combustion chamber of high performance piston engines. Additional polymer or thin film top coatings with dry lubricant ability on the thermally sprayed layers provide a further enhancement of the tribological characteristics. These combined coatings offer low friction and high wear resistance against sliding or oscillating counterparts and are promising candidates for all kind of low friction applications.

CERAMIC MATRIX COMPOSITES (CMC) FOR BRAKE SYSTEMS

The reduction of oscillating masses in the wheel suspension in combination with the thermophysical and chemical stability are the keynotes for the development of ceramic brake disks. The state of the art material, cast iron, needs exceeding sizes to fulfill the challenging requirements concerning the braking performance of modern luxury sports cars[3]. The sizes of cast iron disks induce increasing masses in the wheel suspension thus aggravating the comfort behavior and the driveability problems[4]. SiC as lightweight ceramic material, which provides technical performance concerning the tribological behavior in addition to the temperature and corrosion resistance, could realize both, improved braking performance and comfort behavior[5]. The excellent wear resistance offers the additional possibility for a life time application. But due to the safety sensitiveness of this car component, a sufficient strength and damage tolerance by means of a fiber reinforcement is of greatest importance[6].

Fiber reinforced reaction bonded (RB) ceramics are manufactured by liquid silicon infiltration in porous carbonaceous preforms since more than 25 years[7]. This technique provides very short cycle times in comparison to other densification methods, e.g. chemical vapor infiltration (CVI) or liquid precursor infiltration (LPI) (Fig. 2). The idealized reaction bonding process results from heterogeneous reaction between the liquid silicon and the solid carbon in the matrix, which is produced by thermal degradation of a thermosetting resin binder system during pyrolysis[8]. The main difference regarding the manufacture of these kind of composites is in the preform manufacturing, i.e. the fiber incorporation and forming method[9].

The incorporation of fibers into ceramic matrices increases the effort for the compounding and forming step in comparison to conventional bulk ceramic manufacturing, thus resulting in elevated process cycles and costs[11]. The state of the art forming and curing technologies for the manufacture of carbon fiber reinforced thermosetting based preforms for RB CMC are low pressure processes from the plastics industry, e.g. resin transfer molding (RTM) or autoclave techniques[13]. The pressure molding recently became more and more important, due to the fact, that very short cycle times can be realized (Fig. 2). With respect to the serial manufacturing ability in automotive industry the cycle time for forming and fiber incorporation should be in a range of 5-10 min. This basic requirement can only be fulfilled by pressure molding.

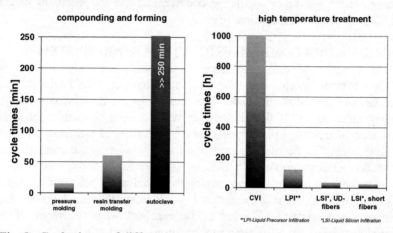

Fig. 2   Cycle times of different processing techniques for fiber reinforced ceramics

For the production of reproducible and homogeneous components controlled die filling during the pressure molding is of greatest importance. Normally fibrous granulates are used but it is also possible to use laminated SMC prepregs (SMC - Sheet Molding Compound) (Fig. 3). In contrast to the limitation on short fibers using granulates, this new approach can also provide the incorporation of long fiber reinforcements and the combination of both, long and short fibers in a great number of variations thus offering the possibility of an exact matching of the structural ceramic properties[9, 14].

SMC is a well established technique in the plastics industry providing a high material flow combined with low process cost[12]. The combination of short and endless fibers is possible due to the in-process fiber cutting. The fibers are deposited together with the resin paste on the carrier film. The final compaction acts as fiber impregnation and results in prepreg sheets. The granulate preparation is a conventional compounding method mostly used in combination with cold pressing in the ceramic industry for manufacturing of bulk materials. An appropriate method for incorporation of fibers is the intensive mixing allowing the homogeneous blending of fibers with powder fillers and binders in combination with granulate formation. The use of thermosetting resin systems as binder requires elevated temperatures for curing during the forming process resulting in short fiber reinforced plastics. The choice of the reinforcing fiber type is strongly limited by the target cost for serially produced brake disks. Only the price of carbon fiber is in an acceptable price range for this application.

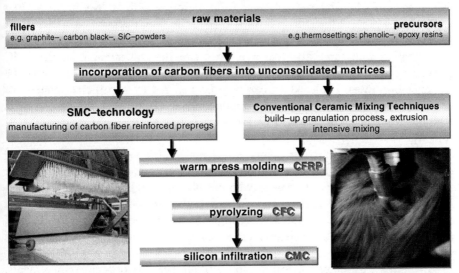

Fig. 3 Appropriate manufacturing technologies for ceramic brake disks with serial production ability

The essential feature for a fiber reinforced ceramic in a structural application is the mechanical behavior, i. e. sufficient strength combined with damage tolerance. The main problem is in realizing adequate mechanical properties using manufacturing technologies with large-scale production ability like the pressure molding. Different SMC- and granulate-based compounds were produced in order to analyze their potential for a brake disk application. Therefore a four point bending test is used operating with the composite testing standard with a torsion arm to sample thickness ratio of more than 10 (ENV 658-3:1992). Fig. 4 shows the stress strain diagram of some selected samples with the corresponding fracture surfaces.

The short fiber reinforced samples made by the granulation process reach the highest flexural strength values up to 220 MPa but in combination with crucial bending strains in the range of 0.13% resulting in a smooth fracture surface without any fiber pull out. The high siliconization rate implicates carbon fiber attack and strong interfaces, so that no energy dissipation occurs. On the other hand the SMC manufactured samples with long fiber reinforcements provide in dependence of the fiber architecture high bending strain values up to 0.33 % with damage tolerant failure behavior and extensive fiber pull out. But this excellent values can only be obtained, if dense CFC preforms with 'self-protecting' fiber architectures prevent the silicon infiltration and the fiber degradation. Using short fibers in the SMC process the values decrease to less than 0.1 due to high siliconization rates and fiber damage. Also the flexural strength of the SMC samples vary strongly and in dependence of the fiber architecture and the siliconization rates. High values can only be realized by a sufficient content of preserved carbon fibers, e.g . the unidirectional (UD) fiber orientation with flexural strengths up to 200 MPa and strain to failure values in the range of 0.35% (Fig. 4).

Other important requirements for a brake disk next to the mechanical properties are a constant high friction coefficient combined with an all weather braking ability. Both aspects can only be fulfilled by a highly siliconized ceramic-like surface with low carbon content and lowest residual porosity [9, 14]. But especially the damage tolerance is ensured by a sufficient number of residual carbon fibers after the siliconization process, thus enhancing the carbon content as well as influencing the residual porosity [10].

There are different possibilities to fulfill all, partially opposite requirements. For the short fiber reinforced components made by granulation the increase in the bending strain values can be performed through protective layers directly on the fiber or on the inner surface of the pyrolysed preform via subsequent infiltration processes. With respect to cost, CVI techniques should be avoided, but liquid resin infiltration or deposition processes can be applied. In the case of the SMC processing 'self-protecting' fiber structures can be used providing structural stability of the component. But these CFC-like structures need oxidation and wear resistant ceramic surfaces, which can be provided by highly siliconized SiC ceramic-like short fiber SMC layers.

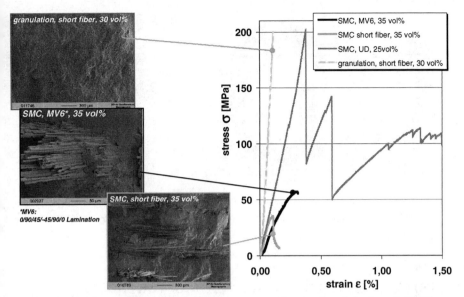

Fig. 4 Mechanical behavior of different manufactured and designed CMCs

This advanced design of the structural ceramic component, the 'sandwich-structure', can be manufactured in one process though appropriate lamination of SMC sheets (Fig. 5).

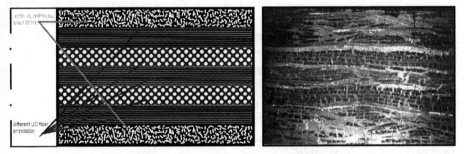

Fig. 5 'Sandwich'-structure consisting of a long fiber reinforced core enveloped in short fiber arrangements (left: schematic, right: cross section)[14, 15]

Both compounding techniques were also used to produce brake disk prototypes in order to determine the technical performance in test stands (Fig. 6).

Fig. 6 CMC brake disk components in comparison
left: short fiber reinforced brake disk in comparison to cast iron and CFC
components
right: SMC manufactured disks with internal ventilation

THERMALLY SPRAYED CERAMIC COATINGS FOR ENGINE APPLICATIONS

Future motor design and engineering require enhanced operation performance and component lifetime, reduced processing time and cost as well as improved environmental and energy efficiencies. Savings in fuel consumption require an optimized engine design and combustion process in combination with a reduction of the total vehicle weight. Nowadays the mass proportion of the engine system concerning the total vehicle weight is in the range of 10 to 15 % [16], [17]. Therefore light weight motor design offers a great potential for a successful mass reduction. State of the art for light metal engines are die-cast aluminum crankcases. Since all parts of a combustion engine interact as a system, all individual components must sustain the combustion pressure and temperature as well as wear and friction effects of moving surfaces in different environmental and lubrication regimes. Due to the worse tribological and thermomechanical operation behaviour of untreated aluminum cylinder liners, a reinforcement of the liner surface is undispensible.

Industrially used protection systems in liner technology are integrated grey cast iron bushings, galvanic coatings based on chromium or nickel, as well as hypereutectic AlSi alloys. Demands to reduce manufacturing complexity, process time and cost on the one side and to improve operation conditions, environmental impacts and recycling management on the other side result in an utilization of thermally sprayed coatings as surface reinforcement for light metal cylinder liners.

To satisfy the combined technical and economical requirements for the highly competitive automobile industry, composites which combine a cheap and easy to machine substrate material, e.g. a light metal alloy, with a high performing coating or surface layer, can be a powerful solution [18], [19], [20]. Especially of tribological interest are coating systems with low friction and wear coefficients. Future engine systems will operate with a lubricant film thickness lower than 1 µm, with biologically degradable fluids, with a life-time lubrication system and must show a reliable operation performance also under mixed and even under dry friction conditions. If traditional liquid lubrication systems cannot be used, the tribological functions must be taken over by material surfaces with solid lubricant capabilities [21]. Bulk ceramic liners with these properties have not been reported.

Thermal spray processes offer the possibility to apply a broad variety of metallurgical, cermet and ceramic coatings on variable material surfaces even with complex geometries and net shape tolerances. For the deposition of internal coatings in light metal cylinder liners the atmospheric plasma spraying APS as well as the high velocity oxygen fuel spraying HVOF can be used. HVOF processes use liquid fuels or fuel gases for high energetic combustion with oxygen ($v_{max} \sim 300 - 600$ m/s; $T_{max} \sim 2,500 - 3,200$ °C). It leads to extremely dense, bulk material like coatings because of the high kinetic impact energy of the hot powder loaded gas jet, as shown in Fig. 7. For the APS process temperatures up to 20,000 °C are obtained, therefore it is mainly used for refractory materials. Due to the reduced kinetic energy of the powder particles in the gas jet ($v_{max} \sim 50 - 100$ m/s), APS coatings have a lamellar, splat like texture, as shown in Fig. 8. During spraying the powder particles are molten in the plasma or autogenous flame and accelerated to the substrate material. The average coating thickness ranges from several microns up to a few millimeters [21]. Precedent to the coating process a roughening and degreasing of the liner surface is performed. The roughening of the surface, e.g. with corundum of defined size, improves the mechanical adhesion of the coating to the substrate material. Following to the deposition process a mechanical honing of the coating surface is performed.

In the case of thermally sprayed internal coatings two principally different processes can be taken into consideration. State of the art technologies for inside coatings are rotating plasma spray devices. The engine block is positioned on a transfer line and the bores are individually coated using a vertically moving, rotating plasma torch [23]. A new process technology of IMTCCC is the deposition of inside coatings with a vertically or elliptically moving APS or HVOF spraying torch and a rotating engine block. The main advantage of this second technique is the much higher deposition rate and improved coating microstructure, e.g. less porosity and better coating adhesion, due to an enhanced jet propagation and thus higher particle velocity and kinetic impact energy with an improved coating densification.

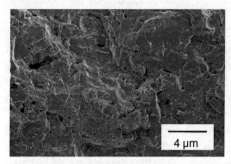 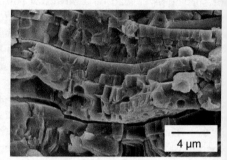

Fig. 7 SEM shot of the fracture cross section of a HVOF sprayed Al₂O₃ coating

Fig. 8 SEM shot of the fracture cross section of an APS sprayed Al₂O₃ coating

An internal coating process under rotation of the engine block needs a manufacturing device which guarantees a reliable and technically safe rotation process as well as a guidance system for the exact positioning of the individual bores under rotation, due to the fact that only the bore positioned in the rotation axis can be coated. The movement and the feed drive of the spraying torch is guided by a multi axis robot system and can be flexibly adapted on bore dimensions and various coating materials. All motion sequences and units, e.g. rotating table, positioning system or robot system are steered by one integral control system, see Fig. 9 and Fig. 10.

Important criteria for a successful implementation of thermally sprayed coatings in light metal cylinder liners are reliable and homogeneous coating qualities in combination with superior surface properties compared to actual technical solutions. All investigated coating systems can be homogeneously deposited over the bore depth. Especially with the HVOF process superior coating qualities with an excellent bonding strength to the substrate material can be reached.

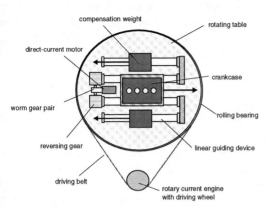

Fig. 9 device for the positioning of the engine block under component rotation

Fig. 10 HVOF equipment for the internal coating process of cylinder bores at IMTCCC

The investigations during the material screening and first bench tests reflect positive results for most of the coating systems in sports car engines. A detailed coating selection can be done after further combustion bench tests in stationary reciprocating engines are finished.

Fig. 11 Coated engine components using the HVOF process

It can be assumed that thermally sprayed coatings will play an important role as protective and functional coatings for light metal engine applications, and will continue to become more and more important in the near future (Fig. 11). They may fulfill what fully bulk ceramic engines once promised.

COMBINED CERAMIC, METALLURGICAL AND CERMET COATINGS WITH DRY LUBRICATION ABILITY ON LIGHT METAL SUBSTRATES

A more widespread use of light metal alloys in tribological applications like guide bars, bearing plates, seat supports or bushings demands powerful functional surface coatings to provide wear protection and to increase the load bearing capacity. Direct contacts of uncoated light metal substrates with sliding or oscillating counterparts result in severe wear, seizing and high friction coefficients, even under lubricated conditions. Due to environmental considerations concerning the effects of lubricants and grease on the ecological systems and in order to decrease the costs for maintenance and service time there is a steady demand for materials and surface coatings with solid lubricant ability and dry friction capability. By the use of these systems either no lubricants at all are needed or the amount of lubricants used can be reduced drastically.

Combined ceramic, metallurgical and cermet coatings on light metal alloys are designed to provide a high wear resistance and high compressive strength for the light metal substrate as well as to exhibit dry lubrication properties. The combined coatings consist of a wear resistant ceramic, metallurgical or cermet primary layer which is applied on the light metal substrate by thermal spraying. Successively, a secondary coating layer with dry lubrication ability is applied to provide a low friction coefficient μ. This secondary layer is either a polymer based lubricating varnish which consists of a polymer matrix containing finely dispersed solid lubricants (e.g. PTFE, MoS_2, graphite, BN_{hex}) or it is applied by thin film technology, i.e. by physical vapor deposition (PVD) or by chemical vapor deposition (CVD). Suitable thin films with dry lubrication ability include diamondlike carbon (DLC) coatings, e.g. deposited by plasma enhanced chemical vapor deposition (PECVD), metal containing diamondlike carbon coatings (Me-DLC), e.g. deposited by reactive physical vapor deposition and molybdenum disulphide (MoS_2) coatings, e.g. deposited by cathode sputtering (DC magnetron sputtering). Due to the limited thermal stability of the light metal alloys it is important that all coating processes are carried out at moderate temperatures well below the melting point of the respective alloys. The fabrication routes of the combined coatings are shown in detail in other publications [25],[26].

In Fig. 12 and Fig. 13 some experimental results are shown demostrating the superb tribological properties of different combined coatings deposited on an aluminum alloy (AlMg3, Fig. 12) and a magnesium alloy (AM50, Fig. 13) in comparison to the respective unprotected light metal surface. The results were obtained using a pin-on-disk tribometer under oscillating sliding movement in unlubricated condition. A 100Cr6 ball was used as counterpart, a normal load of F_N=10 N was applied using an amplitude of osicillation of 5 mm. Without a protective surface coating, the AlMg3 as well as the AM50 substrate show a high and continuously increasing friction coefficient and severe seizing can be observed. The application of an Al_2O_3 primary layer on the AlMg3 substrate by thermal spraying followed by a DLC thin film results in a low friction coefficient in the range of μ=0.1 to μ=0.2. Also the surface properties of the AM50 diecasting can be improved considerably by the deposition of a thermally sprayed TiO_2 coating in combination with a lubricant lacquer or in combination with a DLC or Me-DLC thin film. Even after a possible failure or consumption of the dry lubricant top coat which can be noticed by a sharp increase of the friction coefficient, as visible in Fig. 12 for the combined Al_2O_3 / sputtered MoS_2 coating after 80,000 oscillations, the light metal surface is still protected effectively against wear by the thermally sprayed primary layer. Thus the components can be easily repaired by the replacement of the top coating with dry lubrication ability. A detailed discussion of the experimetal results concerning the tribological behavior of different combined coatings can be found elsewhere [25]-[29].

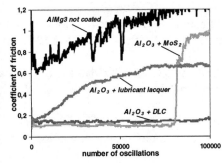

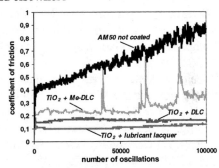

Fig. 12 Coefficient of friction μ vs. number of oscillations of uncoated AlMg3 and different combined coatings on AlMg3 under unlubricated condition

Fig. 13 Coefficient of friction μ vs. number of oscillations of uncoated AM50 and different combined coatings on AM50 under unlubricated condition

CONCLUSIONS

The development of novel solutions and product concepts by composite materials engineering regarding functional coatings on light metal alloys and ceramic matrix composites in combination with advanced manufacturing technologies, process automation and product design opens the way to replace many state-of-the-art bulk material components for different applications in the automotive industry in competitive cost relations. This will result in significant advantages for the customer, e.g. concerning reduction of fuel consumption and waste emissions due to the lower specific weight and an improved comfort behavior and drivability.

REFERENCES

[1] Clark, C.: "Materials and Process form the Vanquish", Automotive Light Metals (7), First Global Media Group (2001), ISSN 1471-6011, pp. 40-42

[2] Hrdliczka, T.: "Der M3 wird Schwere los", AutoMagazin, No. 4, Verlag für Automedien (2002), ISSN 0572-2128, pp. 64 - 67

[3] Jost, K.: "BMW 7 Series: AEI's Best Engineered Vehicle for 2002", Automotive Engineering International, No. 3, Society of Automotive Engineers (2002), ISSN 0098-2571, S. 24 - 32

[4] Kurz, G.; Müller, R.; Fischer, G.: "Bremsanlage und Schlupfregelungssysteme in der neuen Baureihe 5 von BMW", ATZ 98 (4), Vieweg Verlag (1996), ISSN 0001-2785, pp. 188-198

[5] Gadow, R.; Speicher, M.: "Manufacturing and CMC Component Development for Brake Disks in Automotive Applications", Advanced Ceramics and Composites, R. Gadow (ed.), expert-Verlag (2000), ISBN 3-8169-1830-1, pp. 301-312

[6] Speicher, M.: "Wenn Alu und Keramiken Verstärkung brauchen", Serie "Maßgeschneiderte Werkstoffe", Industrieanzeiger (13), Konradin Verlag (2001), ISSN 0019-9036, pp. 50-52

[7] Hillig, W.B. et al.; General Electric Techn. Inform. Serv. 74 CRD 282 (1974)

[8] Fitzer, E.; Gadow, R.: "Fiber reinforced silicon carbide", Am. Ceram. Soc. Bull., Vol. 65 (2), (1986), pp. 326 - 335

[9] Gadow, R.; Speicher, M.: "Manufacturing of Ceramic Matrix Composites for Automotive Applications", Ceramic Transactions, Vol. 128, Advances in Ceramic Matrix Composites VII, eds. P. Bansal et al., The American Ceramic Society (2001), ISBN 1-57498-136-6, pp. 25–41

[10] Gadow, R.; Speicher, M.: "CMC Brake Disks in Serial Production . The Competition between Cost Effectiveness and Technical Performance",

presentation at the 26th International Conference on Advanced Ceramics & Composites, Cocoa Beach, Florida, to be printed in Ceramic Engineering and Science Proceedings, The American Ceramic Society (2002)

11 D.C. Phillips: "Fiber reinforced ceramics", Handbook of Composites, Vol. 4, A. Kelly and S.T. Mileiko (eds.), Elsevier Science Publishers B.V. (1993), ISBN 0 444 864474, pp. 373-428

12 Flemming, M.; Roth, S.; Ziegmann, G.: Faserverbundbauweisen, Springer Verlag (1999), ISBN 3-540-61659-4

13 M. H. Van de Voorde, M. R. Nedele: "CMC`s Research and the Future Potential of CMC`c in Industry", 20th Annual Conference on Composites Advanced Ceramics, Materials and Structures: B, Ceramic Engineering and Science Proceedings (4), (1996), pp. 3–21

14 Gadow, R.; Speicher, M.: „Optimized morphological design for silicon infiltrated microporous carbon preforms", Ceramic Engineering and Science Proceedings 21 [3], The American Ceramic Society (2000), ISSN 0196-6219, pp. 485-492

15 Berreth, K.; Gadow, R.; Speicher, M.: „Fiber reinforced ceramic and method for producing the same", international patent, EP1154970, WO 00/41982,

16 „Wettbewerbsvorsprung durch Spitzentechnologie - Competitive advantages by leading technology"; VDA Verband der Automobilindustrie e.V., Frankfurt am Main, Germany, 2001

17 Niehues J.: „Aluminium-Matrix-Verbundwerkstoffe im Verbrennungsmotor – aluminum matrix composites for combustion engines"; presentation on the MMC workshop, Geesthacht 2000

18 Buchmann M., Gadow R.: "Tribologically optimized ceramic coatings for cylinder liners in advanced combustion engines"; SAE Technical Papers Series, Baltimore 2001, 2001-01-3548, ISSN 0148-7191

19 Buchmann M., Gadow, R.: "Ceramic coatings for cylinder liners in advanced combustion engines, manufacturing process and characterization ", presentation at the 25th International Conference on Advanced Ceramics & Composites, Cocoa Beach, Florida, to be printed in Ceramic Engineering and Science Proceedings, The American Ceramic Society (2001)

20 Buchmann M., Gadow R., Scherer D.: "Ceramic Light metal composites – product development and industrial application" presentation at the 26th International Conference on Advanced Ceramics & Composites, Cocoa Beach, Florida, to be printed in Ceramic Engineering and Science Proceedings, The American Ceramic Society (2002)

21 Woydt M.: „Tribologische Werkstoffkonzepte für den Trockenlauf –

tribological material concepts for dry friction operation conditions"; Tribologie keramischer Werkstoffe, pp. 27 - 43, Expert-Verlag, 2000, ISBN 3-8169-1744-5

[22] Killinger A., Buchmann M.: „Oberflächen wie aus der Pistole geschossen"; Industrieanzeiger No. 8, pp. 46 - 47, E3906C, Konradin Verlag Robert Kohlhammer GmbH, Echterdingen 2000

[23] Barbezat G., Keller S.: "Innovation thrust for car engines"; Sulzer Technical Review 2/96, 1996

[24] Gadow R., Killinger A., Scherer D.: "Ceramic Polymer Composite Coatings", Mat.-wiss. u. Werkstofftechn. 29 (6), pp. 292-299 (1998), WILEY-VCH Verlag GmbH, D-69451 Weinheim, 1998; ISBN 0933-5137

[25] Scherer D., Gadow R., Killinger A.: "Manufacturing and experimental evaluation of combined ceramic polymer coating systems for tribological applications", UTSC, United Thermal Spray Conference 99, 17.-19. März 1999, Düsseldorf; Conference Proceedings; E. Lugscheider, P.A. Kammer (ed.), S. 664-669; ISBN 3-87155-653-X

[26] Gadow R., Scherer D.: "Ceramic and metallurgical composite coatings with advanced tribological properties", Proceedings of 2000 Powder Metallurgy World Congress, 12. – 16.11.2000, Kyoto, Japan; Kosuge, K. and Nagai, H. (ed.), pp. 1112-1115, The Japan Society of Powder and Powder Metallurgy; ISBN 4-9900214-8-7

[27] Friedrich C., Gadow R., Killinger A., Scherer, D.: "Ceramic and metallurgical coatings on magnesium components", 2000 International Conference on Powder Metallurgy & Particulate Materials, 30. Mai – 3. Juni 2000, New York; Advances in Powder Metallurgy and Particulate Materials, pp. 11/63-73, Metal Powder Industries Federation, Princeton, New Jersey, ISBN 1-878954-78-4

[28] Gadow R., Scherer D.: "Composite Coatings on Light Metal Substrates with Dry Lubrication Ability"; E-MRS 2001 Spring Meeting, Strasbourg (France), June 5 - 8, 2001; in Surface & Coatings Technology, 151-152 , pp. 471-477, Elsevier Science (2002); ISSN 0257-8972

[29] Gadow R., Scherer D.: "Functional coating systems on magnesium diecastings for corrosion protection and tribological applications", ATT - Automotive and Transportation Technology Congress and Exhibition, Barcelona, Spain, October 1-4, 2001; ATTCE 2001 Proceedings, Volume 4: Manufacturing, Society of Automotive Engineers, Warrendale, PA, USA, S. 281-286; ISBN 0-7680-0863-8

PROCESSING AND PROPERTIES OF CO-EXTRUDED DIAMOND AND CARBIDE FIBROUS MONOLITHS

Greg Hilmas and Tieshu Huang
University of Missouri-Rolla
Rolla, MO 65409

Zhigang Zak Fang
University of Utah
Salt Lake City, UT 84112

Brian White and Anthony Griffo
Smith Bits, Smith International, Inc.
Houston, TX 77205

ABSTRACT
Functionally designed diamond and cemented carbide architectures were fabricated using the fibrous monolith co-extrusion process and are being considered as a replacement for traditional diamond and cemented carbides for rock drilling bits. The macrostructures of these composites consist of diamond cells with WC(Co) cell boundaries or WC(Co) cells with Co cell boundaries. As with past fibrous monoliths, these new compositions exhibit non-brittle fracture behavior as compared to traditional diamond and cemented carbide monoliths. Cellular structured diamond and WC(Co)/Co fibrous monolith drill bits inserts have been produced by co-extrusion and consolidated by either HTHP (High Temperature High Pressure) or ROC (Rapid Omnidirectional Compaction) processes. The materials show increases in fracture toughness without a significant decrease in wear resistance. Field test results and indentation results for these novel materials will be discussed.

INTRODUCTION
Materials used in petroleum drilling bit inserts require several critical physical and mechanical properties to achieve adequate lifetimes and effective drilling rates. These properties include abrasive wear resistance, chemical wear resistance, high hardness, high temperature strength, high fracture toughness, low thermal coefficient and high thermal conductivity. Typical failure modes for the drill bit inserts include chipping, severe spalling, heat checking and breakage. All of these

failure modes are caused by the brittle behavior of the carbide and polycrystalline diamond materials being used in the industry.

Cook and Gordon[1] first introduced the concept of controlling the crack propagation in brittle materials in the mid-1960's. Functionally designed microstructures became more and more important in toughening brittle materials. Several methods for producing a functionally designed microstructure have been developed in recent years including the following toughening methods:

(1)   Micro-composite methods

- In-situ toughening by creating a desired microstructure through control of grain growth and grain boundary phase development[2,3]. A microstructure with elongated grains and a weak grain boundary phase will provide enhanced fracture toughness. Propagating cracks will be deflected by the elongated grains and propagate in weak grain boundary phases. This increases the energy required to continue crack propagation, providing resistance-curve behavior. However, this method is limited by the materials systems that can produce an elongated grain and weak grain boundary phases.

- Fiber reinforcing can be used to add a higher strength second phase. Deflection of propagating cracks along with effects of fiber "pull-out" forces are the main reasons for providing enhanced toughness[4]. The disadvantage of this method is the cost, both for the fibers and the added cost of complex fabrication techniques.

(2)   Macro-composite methods

- Fibrous monolithic structures can be developed by extruding coaxial feedstock consisting of powder/polymer blends of ceramic, metal or cermet powders. The division of strong but brittle materials into isolated cells separated by relatively weaker or ductile materials into the surrounding cell boundaries produce a cellular structure that is truly engineered[5-8]. Si_3N_4-BN and SiC-BN fibrous monoliths were the most successful examples in the early stages of this technology[9,10].

Fibrous monolithic structures are distinctive, functionally designed architectures that can be produced without a significant added cost. The most important engineering advantage of the process is its capability to provide some of the latter toughening effects in most materials systems without significantly decreasing other mechanical properties. In other words, in systems such as the carbide/metal system[11] for petroleum drill bit inserts, as reported in reference 11, the engineer can avoid the classic trade-off of gaining toughness only to lose wear resistance.

EXPERIMENTAL PROCEDURE

1. Fibrous Monolith Fabrication

Functionally designed, fibrous monoliths of polycrystalline diamond/cemented tungsten carbide (PCD/WC(Co)) and cemented tungsten carbide/cobalt metal (WC(Co)/Co) were produced in this study by blending the constituents with organic polymers and co-extruding them as a core/shell structure from controlled geometry feed rods. The architecture of the fibrous monoliths in this study were either PCD cells with a WC(Co) cell boundary or WC(Co) cells with a Co cell boundary. The procedures for mixing the powders with organic binders and fabricating the fibrous monolith filaments by multiple co-extrusion to achieve the cell/cell boundary architecture have been described in detail elsewhere[6,8,10].

The PCD powder was grade D1 (diamond particle size: 2-12μm, Co content: 5-20wt% and was obtained from Smith International, Inc., Houston, TX). The cemented tungsten carbide powder was grade WC2 (WC particle size: 2-12 μm, Co content: 5-20 wt% and was obtained from Smith International, Inc.). The Co powder was a −325 mesh (powder from Cerac, Milwaukee, WI).

2. Specimen Fabrication and Testing

For the D1/WC2 (referred to cellular PCD or C-PCD) materials, a final feed stock was produced from multi-cellular filaments by bundling and laminating the filaments in a 21.25mm diameter cylindrical die at 140-160 °C and 20 MPa. From this feed stock disks were sliced and used as the wear surface for drill bit inserts for petroleum drilling. These components were debinded and pressed using HTHP process at 6 GPa and 1400 °C by Smith Megadiamond (Provo, UT) for drill rig testing.

For the WC2/Co materials, a feed stock in 21.25mm diameter and 64mm long was produced by laying up multi-cellular filaments in a 21.25mm cylinder die and warm laminating them at 140-160 °C and 20 MPa. From these feedrods disks were sliced. A multi-step debinding and pre-sintering was carried out in Ar + H atmosphere (under flowing gas) at temperatures up to 1000 °C. The pre-sintered samples were sintered by ROC at 800 MPa and 1240 °C.

Several field tests using the diamond-based fibrous monoliths were carried out in different parts of the world, including East Texas, Wyoming; Eastern Kentucky; Kenai Peninsula, Alaska; Ross Field, North Sea, UK; Lennox, UK; Gulf of Mexico; Belayim Land, Egypt; and in Kuwait. Cellular drill bit inserts from tested drill bits were then analyzed for damage by optical and scanning electron microscopy. Indentation tests were carried out on polished surfaces of WC2/Co using a LECO DM-400 Hardness Tester. The load was 50 kg.

RESULTS AND DISCUSION

1. Microstructures

Figure 1 shows the microstructure of (C-PCD) material. The PCD D1 diamond cells were uniformly surrounded by WC2 cell boundaries. This type of engineered microstructure is desirable for improving the fracture toughness of brittle materials. The architecture matches the predicted design with an average cell size of ~200 μm and a cell boundary thickness of ~32 μm as shown in Figure 1 (a). Figure 1 (b) is a side view showing the cross-section of a cellular drill bit insert. The structure of the insert is a multi-layer design with the cellular D1/WC2 as the wear surface layer on the right side of the figure.

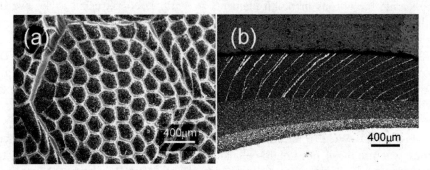

Figure 1. SEM micrographs of C-PCD: (a) top view; (b) side view, multi-layer structure, the right layer is C-PCD

Figure 2 shows the microstructure of the WC2/Co cellular materials. The micrograph clearly shows the designed cellular architecture with isolated WC2 cells and distinct cell boundaries of Co material (~25 vol%). The results demonstrated that the ROC sintering process is suitable for densifying the structure to >99% of theoretical density. Since the sintering was performed at a temperature of 1240 °C, much lower than the melting point of cobalt (~1495 °C),

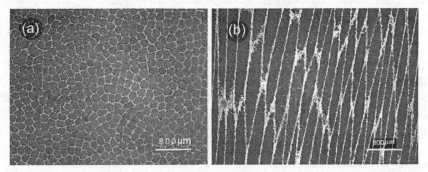

Figure 2. SEM micrographs of the cellular WC2/Co, (a) top and (b) side view

the sintering is solid-state. Figure 2 (a) shows the top view of the cellular structure with WC2 cells that are ~200 μm in size with Co cell boundaries that range from 20-35 μm in thickness. Figure 2 (b) shows the side view of the cellular structure. As with the PCD-based co-extruded materials, a uniform architecture of cells/cell boundaries has been achieved.

2. Field test results

Figure 3 shows a SEM image of a field-tested rock bit insert containing C-PCD materials. It can be seen from the image that the cellular structure interrupts cracking and chipping propagation. While small pieces of cell structure were chipped off during the testing the cracking and chipping were isolated to a limited area.

Figure 4 provides a comparison of standard PCD layered bit inserts with the C-PCD layered bit inserts after field drilling tests. The results show the C-PCD layered bit inserts possess higher toughness and resistance to cracking and catastrophic damage. Figure 4 (a) shows successful C-PCD inserts that were run in a field test as alternating bit inserts with standard PCD inserts on a roller-cone drill bit. The C-PCD inserts all survived without catastrophic damage while the standard PCD inserts had all failed (Figure 4 (b)).

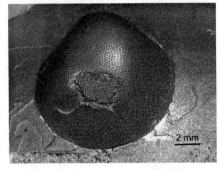

Figure 3 SEM image a used rock bit insert with C-PCD material

Table 1. Cellular vs. Standard Drill Bit Performance Comparison

Drill Bit	Footage (ft)	ROP (fph)	Hours	Krevs
C-PCD Bit[*]	337	11.4	29.5	415
Standard PCD Bit	123	13	11.1	110

*Drill Test #1

Figure 5 shows the performance of a roller-cone bit containing C-PCD inserts compared to that of the "Median" data for several runs of standard PCD containing roller-cone bits. These bit comparisons were run in successive field tests in the Kenai Peninsula, Alaska. The results show the substantial performance gains for the C-PCD containing bits compared to the standard bits. The actual field record is shown in Table 1. The table compares the total feet drilled (footage), the rate of penetration (ROP), the hours of drilling time (in feet per hour) and the total number of bit revolutions (in kilo-revolutions). From Table 1, the C-PCD bit has a slightly lower value ROP, but the total number of feet drilled, the hours of drilling, and the total bit revolutions are all 250-350% higher than that of standard PCD containing (Median) bits.

Figure 4. SEM micrographs of standard PCD bit inserts and C-PCD bit inserts after field tests: (a) is a C-PCD layered bit insert and (b) is a standard PCD layered bit insert

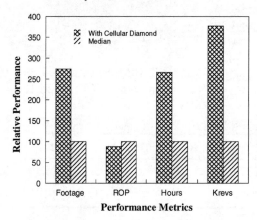

Figure 5 Performance of C-PCD bits in comparison to that of several standard (Median) bits in drill test #1

Figure 6 shows another performance comparison of C-PCD bits with typical (Median) data from standard PCD bits tested in Belayim Land, Egypt. Table 2 shows the field record for C-PCD bit vs. standard PCD bits for the same performance metrics. The overall performance of C-PCD bit is again higher for all performance metrics compared to the standard PCD bits.

The better performance in these applications comes from increased toughness and wear resistance. Better toughness and wear resistance comes from functionally designed microstructures. The functionally designed microstructures improve toughness and wear resistance in the following three ways. First, the large hard granules in the cells provide sufficient wear resistance and determine the overall wear rate. Second, the more ductile interpenetrating cell boundary provides fracture toughness, chipping resistance. Third the cellular microstructure is capable of functionally interrupting cracks.

Table 2. Cellular vs. Standard Drill Bit Performance Comparison

Drill Bit	Footage (ft)	ROP (fph)	Hours	Krevs
C-PCD Bit*	1161	21.3	54.5	555
Standard PCD Bit	794	12.9	52	248

*Drill Test #2

1. Crack initiation and propagation

Figure 7 shows the microstructures of an indented WC2/Co cellular composite. The indent was produced with Vickers diamond indenter and a 50 kg load. Radial/median cracks from the corners of the indents, initiating either in the cells or the cell boundaries, propagated to the cell boundaries. When the cracks reached the cell boundaries they would change their propagation direction (deflect) in order to remain along the cell boundaries. Since the cell boundaries are ductile materials, the cracks only propagate for short distances (<400μm). The crack energy is dissipated over distances as short as a few cells. Due to the resistance curve behavior of this material, fracture toughness measurements could not be made using an indentation technique.

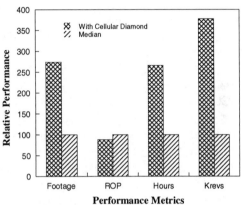

Figure 6 Performance of C-PCD bits comparison to that of Median bits in drill test #2

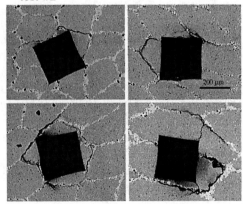

Figure 7 Microstructures of indented WC2/Co cellular structured sample

CONCLUSION

Cellular PCD (D1/WC2) and cellular WC2/Co materials were

successfully designed and developed. These functionally designed composites possess high toughness and wear resistance. Petroleum drilling field tests using cellular PCD drill bit inserts demonstrated significant performance gains versus standard PCD containing drill bits. The cellular structure can effectively increase the overall fracture toughness and wear resistance of hard, polycrystalline diamond and carbide materials without decreasing wearing resistance significantly.

References

[1]J. Cook and J. E. Gordon, "A mechanism for the Control of Crack Propagation in All-Brittle System," *Proc. R. Soc. London*, 282, 508-20 (1964).

[2]P. F. Becher, "Microstructural Design of Toughened Ceramics," *J. Am. Ceram. Soc.* **74** [2] 255-66 (1991)

[3]N. T. Padture, "In Situ-Toughened Silicon Carbide," *J. Am. Ceram. Soc.* **77**[2] 519-25 (1994)

[4]P. F. Becker, "Recent Advances in Whisker-Reinforced Ceramics," pp 179-95 in *Annual Review of Materials Science*, Vol. 20 Edited by R. A. Huggins. Annual Review, Palo Alto, CA (1990)

[5]G. Hilmas, A. Brady, U. Abdali, G. Zywicki, and J. Halloran, "Fibrous Monoliths: Non-Brittle Fracture from Powder Processed Ceramics," *Mat. Sci. Eng. A* A915, 263 (1995).

[6]D. Popovic, J. W. Halloran, G. E. Hilmas et al, "Process for Preparing Textured Ceramic Composites," U.S. Pat. No. 5645781, July 8, 1997

[7]D. Kovar, B. H. King, R. W. Trice, and J. W. Halloran, " Fibrous Monolithc Ceramics," *J. Am. Ceram. Soc.* **80** [10] 2471-87 (1997).

[8]D. Kovar, G. A. Brady, M. D. Thouless, and J. W. Halloran, "Interfacial Fracture between Boron Nitride and Silicon Nitride and its Applications to The Failure Behavior of Fibrous Monolithic Ceramics," *Mat. Res. Soc. Symp. Proc.* Vol. 409 243-248 (1996) Materials Research Society.

[9]G. E. Hilmas, G. A. Brady, and J. W. Halloran, "SiC and Si_3N_4 Fibrous Monoliths: Nonbrittle Fracture from Powder-Processed Ceramics Produced by Coextrusion," pp. 609-14 in *Ceramic Transactions*, Vol. 51, Fifth International Conference on Ceramic Processing Science and Technology. Edited by H. Hausner, G. L. Messing, and S.-I. Hirano. American Ceramic Society, Westerville, OH 1994.

[10]A. Brady, G. E. Hilmas, and J. W. Halloran, "Forming Textured Ceramics by Multiple Coextrusion" *ibid* pp. 297-301

[11] Z. Zak Fang, A. Griffo, B. White, G. Lockwood, D. Belnap, G. Hilmas, and J. Bitler, *International Journal of Refractory Metals and Hard Materials* 19 (2001) 453-459

LOW - TEMPERATURE INFILTRATION OF BORON CARBIDE - ALUMINUM COMPOSITES

Fuhong Zhang, Kevin P. Trumble and Keith J. Bowman
School of Materials Engineering
Purdue University
West Lafayette, IN 47907

ABSTRACT
Boron carbide and aluminum can form many different reaction products during infiltration. Suppressing the reactions might lead to better performance of composites. However, spontaneous infiltration only takes place when the temperature is high enough (typically >1200°C) to obtain a low contact angle. A low - temperature infiltration scheme was designed and full infiltration of boron carbide by aluminum at T<1000°C was achieved. Microstructure characterization revealed significantly less reaction products than previously reported.

INTRODUCTION
Ceramic armor has been developed for many years. Many different ceramics have been used as armor materials. Ceramic armor tiles, typically alumina (Al_2O_3), silicon carbide (SiC), silicon nitride (Si_3N_4) and boron carbide (B_4C), are presently being fabricated commercially and are primarily used to blunt armor piercing projectiles.

However, the imperative for enhanced mobility, increased survivability and reduced logistical burden for both personnel and vehicles demands advanced materials with better performance. Many studies have been conducted in an attempt to identify a high-performance, lightweight armor material that allows for increased speed and mobility.

Among all these ceramic armor materials, boron carbide is well known for its high hardness and low density. However, it suffers from low fracture toughness. When it is combined with lightweight metal, such as aluminum, to produce a metal - ceramic composite with improved fracture performance, it is a very promising material in the situations where high stiffness, high hardness, wear resistance and light weight are demanded.

Development of boron carbide–aluminum composites has been reported by many researchers. [1,2,3,4] Liquid aluminum infiltration methods are most commonly used in the preparation. Chemical reactivity between aluminum and boron carbide, and the wetting angle variation with temperature and processing time have also been investigated. [5,6] It is reported that Al and B_4C are highly reactive at elevated temperatures and that the contact angle drops dramatically at short time with increasing temperature under vacuum and inert gas atmospheres. Many different reaction products can be present in the composite system (Table 1). In most cases, these products are unwanted because of their lower hardness and stiffness than those of boron carbide and higher brittleness than that of aluminum. Reducing the infiltration temperature might result in composites with fewer reaction products. However, spontaneous infiltration is only possible when the temperature is high enough that a low contact angle and wetting conditions are established. [4,6]

In the present work, infiltration processes were investigated and a low temperature infiltration process was developed to ensure a minimum amount of reaction products in boron carbide – aluminum composites.

Table 1: Common reported phases in boron carbide – aluminum composites.

Phases	Structure	Vicker's Hardness (test load) (kg/mm^2)
B_4C	Rhomobohedral	3250-3350 (1kg)
AlB_{10} ($Al_4B_{24}C_4$)	Orthorhombic	2840(20g)
Al_3BC	Hexagonal	1400 (10-20g)
Al_4C_3	Rhombohedral	1230(300g)
AlB_2	Hexagonal	1050(10g)
AlB_{12}	Tetragonal	-
β-AlB_{12}	Tetragonal	-
Al	FCC	19 (100g)

EXPERIMENTAL PROCEDURE

Ceramic preforms were prepared by centrifugal casting as previously reported. [4] Boron carbide powder (BO-301, 1500 grit, Atlantic Equipment Engrs., NJ, USA) was mixed with de-ionized water under stirring to form a slurry of 20 vol% solid content, followed by ultrasonic treatment to achieve homogeneity. The pH was adjusted to about 7.6 to 8.0 using NH_4OH. Boron carbide green bodies were cast layer-by-layer by using a floor mounted centrifuge (Model RC 3C plus, Sorvall®, Newton, CT). After casting each layer, the top supernatant was removed and slurry was added to cast next layer. A centrifugal acceleration force of ~2000 times standard gravity was used in casting. Drying was carried out at room temperature for 24 hours before the casting was removed from the bucket and finally dried in an oven at 60°C until completely dry. A typical casting is about 14 mm thick and 38 mm in diameter, with each layer 600 to 700 μm in thickness. The dried bodies were then hot-pressed uni-axially at 1800-1900°C/10 MPa to partially densify and obtain an interconnected boron carbide network with a layered structure.

The preforms were then infiltrated with molten aluminum (99.99% pure shot, Alcoa Specialty Metals, Banton, NC). Infiltrations were carried out using two different methods. (1) Spontaneous infiltration process: A boron carbide preform was put into a crucible and aluminum shot was loaded on top of the preform. Then infiltration was conducted under a rough vacuum (<100μTorr) in a temperature range of 1200 to 1300°C, at which the contact angle could drop to a small enough value to let spontaneous infiltration take place. A fast ramp rate, ~30°C/min, was used to minimize the reactions between boron carbide and aluminum. (2) Low temperature infiltration process: Based on the observation that the oxide skin exists on the aluminum surface and it only disappears at close to 1200°C under rough vacuum, it might play a role in wetting by separating aluminum from direct contact with the boron carbide. It might be possible to lower the infiltration temperature by promoting a fresh contact between aluminum and boron carbide. A schematic setup for low temperature infiltration process is shown in Figure 1 and the temperature profile is shown in Figure 2. When the temperature increases and aluminum melts under vacuum, the high contact angle between aluminum and alumina prevents molten aluminum from flowing through the stacking pores due to the capillary pressure. The alumina bead bed separates the aluminum and boron carbide from direct contact during heating, and thus avoids the chemical reactions at high temperature. When temperature was above 1200°C, the molten aluminum would form a gas tight "molten aluminum seal" over the alumina beads. Then the temperature was lowered to the infiltration temperature. After reaching a stable temperature, argon/hydrogen gas was released into the furnace chamber up to a pressure of 1 atm. forcing the molten aluminum through the pores between alumina beads. Direct contact between boron carbide and molten aluminum was established. Infiltration was conducted, followed by cooling to room temperature. The critical pressure needed to achieve the flow of aluminum through the alumina bed is dependent on the size the alumina beads or powder. It is inversely proportional to granule size. The size of the beads must be chosen so that the molten aluminum does not flow through the alumina bead bed by gravity.

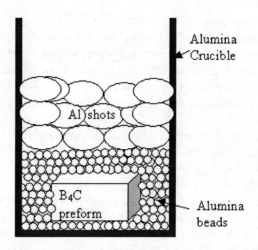

Figure 1: Schematic setup for low temperature infiltration of Al into boron carbide

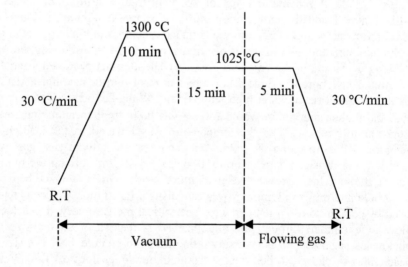

Figure 2: A temperature – pressure profile for low temperature infiltration

After infiltration, the composite samples were cut and polished using SiC, diamond and colloidal silica abrasives. Then the phase content was evaluated by XRD and quantitative image analysis and hardness measurements were carried out.

RESULTS AND DISCUSSION

Microstructure of the graded layered composites

Optical micrographs taken from a cross section of the boron carbide-aluminum composite are shown in Figure 3. Two layers of the composites are shown at a lower magnification in the background. The ceramic and metal phase network can be observed. The multilayered structure and the variation of the particle size within each layer are clearly shown from the four higher magnification micrographs of the selected areas along a trans-layer direction. The gradient in particle size is a characteristic of the centrifugal casting processing. During centrifugal casting, the large particles settle faster than small particles, preferentially depositing on the bottom of the each layer. The particle size variation, measured by quantitative stereology, is shown in Figure 4. The particle size varies gradually from ~2 μm in side to ~10 μm in other side of each layer. In contrast to the gradient in boron carbide particle size across the layers, the volume fraction of boron carbide is similar throughout the composite, suggesting that its packing is uniform and independent of the particle size in this process.

The particle size distribution has a direct effect on the amount of reaction product. The extent of chemical reactions inversely related to particle size. The smaller the particle size, the more reaction product can be observed. The aluminum content at different position of a layer is present in Figure 5, where the x-axis is the normalized position by the layer thickness. It can be seen that the aluminum content decreases with time and it decreases faster in the fine particle region compared to the coarse particle region. Figure 6 shows the peak intensity for one chemical reaction product, Al_3BC. It was measured in one composite layer by cutting a taper and using small spot X-ray beam. The peak intensity is larger for the fine particle region and smaller for the coarse particle region. This observation is consistent with Figure 5, where the fine particle region has more reaction products and less aluminum.

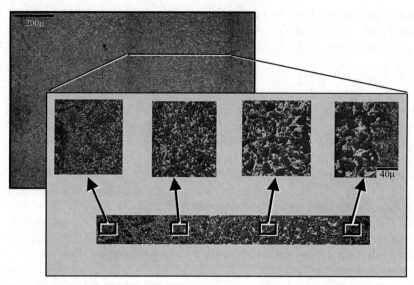

Figure 3: Microstructure of the layer boron carbide-aluminum composite.

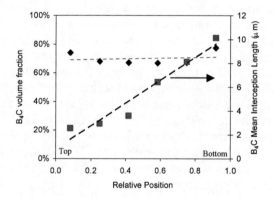

Figure 4: Boron carbide particle size distribution and volume fraction across a single layer of composite.

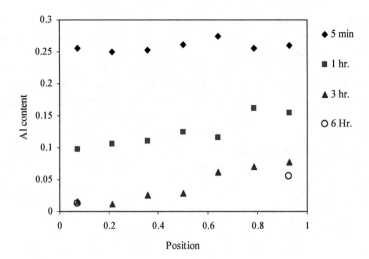

Figure 5: Aluminum content within a layer after processing at 1250°C for different times.

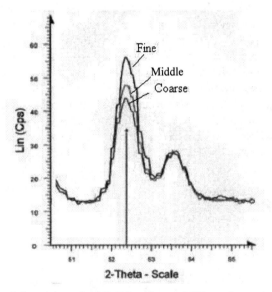

Figure 6: Peak intensity of Al_3BC for fine, middle and coarse region inside one composite layer.

Spontaneous Infiltration

It was observed that at ~1200°C under rough vacuum (~13 Pa (100 mTorr)), aluminum on the top of a porous boron carbide preform start to spread over and spontaneously infiltrate the preform in a few minutes. As reported by Halverson,[6] aluminum spontaneously infiltrates boron carbide when the temperature is high enough to promote a decreased contact angle. Spontaneous infiltration was not observed under flowing argon atmosphere, even at higher temperatures. This result is consistent with Halverson's measurement of contact angle change in flowing argon, where even after 100 minutes at 1200°C the contact angle was still greater than 70°.[6] From calculations for close-packed spheres, Hilden[7] and Trumble[8] found that contact angle of ~50° or less is necessary for spontaneous (pressureless) infiltration. It is worth noting that the oxide skin on the aluminum may play a role in keeping it from direct contact with the boron carbide. During heating, the aluminum shots kept its original shape until the temperature was above 1050°C, which is much higher than the melting point of aluminum.

The effects of processing time and temperature on phase content of the composite were investigated. High temperature and long process time help drop the contact angle and promote spontaneous infiltration, but they also lead to more reaction products, which can adversely affect the composite mechanical properties. Figure 7 shows X-ray diffraction patterns of composites produced at 1200°C for 160 minutes and 20 minutes. After 160 minutes, many different kinds of reaction products were found whereas after 20 minutes no phase other than boron carbide and aluminum was detected. This observation is also consistent with aluminum content in the composites (Figure 8). With longer processing times, the residual aluminum content decreases. However, even after a long processing time, the rate of the consumption of aluminum slows down and a small amount of aluminum remains present in the composites. This is consistent with the observation by Viala and coworkers[5] that reaction product Al_3BC can form a passive layer and serve as a reaction barrier, which slows down reaction rate and makes completely depletion of aluminum very difficult.

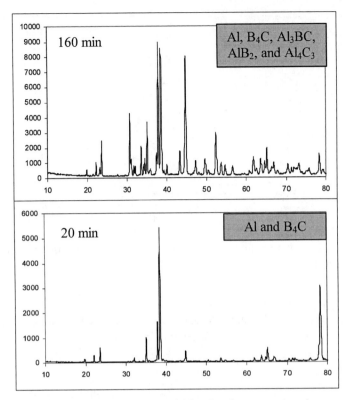

Figure 7: XRD pattern for boron carbide-aluminum composites processed at 1200°C for different times.

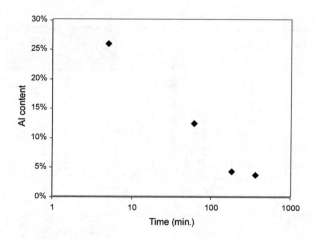

Figure 8: The average aluminum content variation with processing time at vacuum at 1250°C.

Low Temperature Infiltration

Low temperature infiltration experiments were carried out over a temperature range of 800 – 1050 °C. The result is shown in Table 2.

Table 2: Results of low temperature infiltration.

Temperature (°C)	1050	975	900	800
Observations	Fully infiltrated	Fully infiltrated	Partially infiltrated	Not infiltrated.

It was observed that the average content of reaction products reduced significantly from 6.1% when infiltration was processed at 1250°C for 5 min to less than 1.5% when it was processed at 1025°C for 5 min and less than 1.0% when it was processed at 975°C for 5 min by using low temperature infiltration process. Figure 9 shows the microstructures of samples infiltrated at 1250°C and 975 °C, where the latter have fewer reaction products.

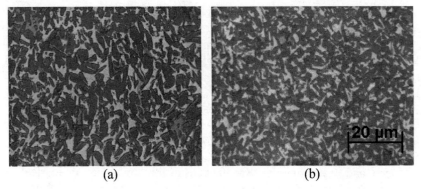

(a)   (b)

Figure 9: Infiltration (a) at 1250°C for 5 min, (b) at 975°C for 5 min.

SUMMARY

Graded multilayer boron carbide composites were successfully prepared by spontaneous infiltration and low temperature infiltration process, respectively. Effects of infiltration conditions on phase content were investigated. Centrifugal casting was used to produce the particle size gradient within each layer. The amount of reaction products is inversely related to particle size and the fine particle regions have more reaction products that the coarse particle region. Long processing time leads to more reaction product and greater aluminum consumption.

A low temperature infiltration process was developed, which can significantly reduce the amount of reaction products from 6.1% when infiltration was processed at 1250°C for 5 min to less than 1.5% when it was processed at 1025°C for 5 min and less than 1.0% when it was processed at 975°C for 5 min.

ACKNOWLEDGEMENT

This research is supported by the United States Army Research Office MURI grant No. DAAH04-96-1-0331.

REFERENCES

[1] J. J. Petrovic, K. J. McClellan, C. D. Kise, R. C. Hoover, and Scarborough, "Functionally Graded Boron Carbide," MRS 1997 meeting, Boston, MA, 1-5 Dec., 1997

[2] A. J. Pyzik. and D. R. Beaman, "Al-B-C Phase Development and Effects on Mechanical Properties of B_4C/Al-Derived Composites", J. Am. Ceram. Soc., 78 [2] 305-12, 1995.

[3] A. J. Pyzik, U. V. Deshmukh, S. D. Dunmead, J. J.Ott, T. L. Allen, and H. E. Rossow, US patent 5521016.

[4] F. Zhang, K. P. Trumble and K. J. Bowman, "Graded Multilayer Boron Carbide-Aluminum Composites," 6th International Symposium on Functionally Graded Materials, Estes Park, Colo. Sep. 2000.

[5] J. C. Viala, J. Bouix, G. Gonzalez, and C. Esnouf, "Chemical Reactivity of Aluminum with Boron Carbide," J. Mater. Sci. 32 (1997) 4559-4573.

[6] D. C. Halverson, A.J. Pyzik, and I. A. Askay, "Processing and Microstructure Characterizaton of B_4C-Al Cermets," *Ceram. Eng. & Sci. Proc.*, 6, [7-8], 736-741, 1985

[7] J. Hilden, and K. P. Trumble, "Spontaneous Infiltration of Non-cylindrical Porosity: Large Pores," Materials Science Forum, Vols 308-311 (1999) p127-162.

[8] K. P. Trumble, "Spontaneous Infiltration of Non-Cylindrical Porosity: Close-Packed Spheres," Acta Mater. 46, [7], 2363-2367, 1998.

HARD AND TOUGH ZrO$_2$-WC COMPOSITES FROM NANOSIZED POWDERS

G. Anné, S. Put, J. Vleugels and O. Van Der Biest

Department of Metallurgy and Materials Engineering
Katholieke Universiteit Leuven
Kasteelpark Arenberg 44
B-3001, Heverlee, Belgium

ABSTRACT
    Yttria-stabilized zirconia based composites with a tungsten carbide content up to 50 volume percent were prepared from nanopowders by means of hot pressing. The mechanical properties were investigated as a function of the WC content. The hardness increased from 12.3 GPa for the Y-TZP ceramic up to 16.4 GPa for the composite with 50 vol. % WC. An optimum fracture toughness of 9 MPa m$^{1/2}$ was obtained for a 40 vol. % WC composite. The hardness and fracture toughness of the composites with a nanosized WC source were significantly higher than with micron-sized WC. Transformation toughening was found to be the major toughening mechanism in the ZrO$_2$-WC composites.

INTRODUCTION
    Tetragonal zirconia polycristalline (TZP) materials have excellent mechanical properties such as a high bending strength and excellent fracture toughness, due to transformation toughening [1]. The modest hardness of these materials however limits their use for wear applications.
    During the last decade, the applicability of zirconia to induce transformation toughening of non-oxide materials by the martensitic stress-induced transformation of tetragonal ZrO$_2$ (t-ZrO$_2$) to monoclinic ZrO$_2$ (m-ZrO$_2$) was investigated. Amongst the composite systems investigated are mullite-ZrO$_2$ [2], Si$_3$N$_4$-ZrO$_2$ [3], TiB$_2$-ZrO$_2$ [4,5,6,7] and Ti(C,N)-ZrO$_2$ [8,9]. Other reported systems in which the ZrO$_2$ is present as the matrix phase includes ZrO$_2$-Al$_2$O$_3$-TiC [10], ZrO$_2$-SiC [11], ZrO$_2$-Cr$_2$O$_3$, -Cr$_3$C$_2$ and -Cr$_7$C$_3$ [12]. Recently, ZrO$_2$-based composites with hard TiB$_2$, TiCN, TiN and TiC secondary phase additions were described [13]. Although excellent toughness values could be achieved with an

---

yttria-coated ZrO_2 starting powder, the increase in hardness aimed at by the addition of the hard secondary phases was rather modest [13]. The addition of micron-sized WC grains to a ZrO_2 matrix was reported to increase the hardness significantly, although the reported toughness values were rather low [14,15]. The major toughening mechanisms in the ZrO_2-WC composites were identified to be microcracking, crack bridging, crack deflection and crack branching [16].

In this paper, ZrO_2-WC composites were prepared from nanopowders. Their mechanical properties are evaluated and a route to optimise the fracture toughness of the composites is described. Moreover, the contribution of the individual toughening mechanisms to the total toughness of the composites is elucidated.

EXPERIMENTAL PROCEDURE

Two WC powder sources were used. Eurotungstène grade CW5000 with an average particle size of 0.8-1 µm and Nanodyne grade Nanocarb WC/Co12. The latter starting powder consists of about 50 µm large WC/Co composite agglomerates. The individual WC crystal size however is claimed to be between 20 and 40 nm. The Co was leached in HCl at 50°C. After leaching, the powder was washed with distilled water to remove the chlorides. Final washing was done in iso-propanol to obtain soft WC agglomerates. The washed powder was subsequently dried at 90°C for 12 hours. The effect of the cobalt leaching and additional milling with WC/Co balls in iso-propanol for 48 hours on the Nanocarb WC/Co 12 are illustrated in Fig. 1. The agglomerates of the original composite powder disintegrated in finer agglomerates due to the removal of the Co binder. Subsequent Turbula mixing in iso-propanol using hardmetal milling balls ($\varnothing$ = 4-5 mm, Céramétal grade MG15) for 48 hours resulted in submicron nanoparticle agglomerates.

The ZrO_2 powders are commercially available powders, i.e., pure monoclinic (Tosoh grade TZ-0) and 3 mol % Y_2O_3 co-precipitated (Daiichi grade HSY-3U) ZrO_2 nanopowders. The crystal size of both powders is around 30 nm. Powder mixtures with an overall Y_2O_3-content between 1.75 and 3.00 mol % were obtained by mixing both ZrO_2 starting powders in the appropriate ratio.

For the preparation of the composites, 60 g of powder was mixed on a multidirectional mixer (type Turbula) in iso-propanol with pure alumina balls to break the agglomerates during mixing. After mixing, the iso-propanol was removed and the dry powder mixture was hot pressed (W100/150-2200-50LAX, FCT Systeme der Strukturkeramik GmbH, Rauenstein, Germany) in vacuum ($\approx$ 0.1 Pa) under a mechanical load of 28 MPa. The samples were separated from the furnace atmosphere by the graphite hot press set-up, whereas the sliding contacts were sealed by boron nitride. The ceramic samples were hot pressed for one hour at 1450°C, with a heating rate of 50°C/min and a cooling rate of 10°C/min.

Fig. 1. Original Nanocarb WC/Co powder agglomerates (a) and WC grains after Co-leaching and mixing for 24 hours in iso-propanol (b).

The Vickers hardness HV_{10} was measured on a Zwick hardness tester (model 3202, Zwick, Ulm, Germany) with an indentation load of 10 kg. The reported values are the mean of at least 5 indentations. The indentation fracture toughness, K_{IC}, was calculated from the length of the radial cracks. The reported values are the mean of at least 5 indentations. The fracture toughness was calculated according to the formula of Anstis et al [17]. The elastic modulus, E, of the ceramic specimens was measured by the resonance frequency method [18]. The resonance frequency was measured by the impulse excitation technique (Grindo-Sonic, J.W. Lemmens N.V., Leuven, Belgium). The density of the specimens was measured in ethanol, according to the Archimedes method.

For microstructural investigation, a SEM (XL30FEG, FEI Company, Eindhoven, The Netherlands) was used. X-ray diffraction analysis (XRD, 3003-TT, Seifert, Ahrensburg, Germany) using Cu $K\alpha$ radiation (40 kV, 30mA) was used for phase identification and determination of the relative phase contents. For the calculation of the monoclinic and tetragonal phase content, the method of Toraya was used [19].

For selected composites, a comparison of the transformability of the tetragonal zirconia, defined as the fraction that can be transformed into monoclinic ZrO_2, was obtained by measuring the difference in monoclinic phase content on polished and fractured surfaces, obtained by breaking samples in a 3-point bending test set-up.

RESULTS

Microstructural investigation of the hot pressed ZrO_2-WC composites revealed that full densification could be achieved after 1 hour at 1450°C. Homogeneous ZrO_2-WC microstructures were obtained after 48 hours of mixing. Some representative backscattered electron micrographs are shown in Fig. 2.

Three phases can be clearly differentiated, i.e. ZrO_2 (grey), WC (white) and Al_2O_3 (black). The Al_2O_3 grains originated from the alumina mixing medium. Although grain growth of the WC phase during sintering is obvious, the final grain size of the WC phase in the Nanocarb powder based composites is substantially smaller than that in the micron-sized powder based. The angular geometry of the WC grains remained after sintering, indicating that the ZrO_2 and WC phases are chemical compatible at the sintering temperature. The maximum grain size of the WC grains in the Nanodyne WC powder based ZrO_2-WC (80/20) composite is below 1 µm. At WC contents above 20 vol %, the WC grains tend to agglomerate in small clusters. WC network formation is obvious at WC contents above 40 %.

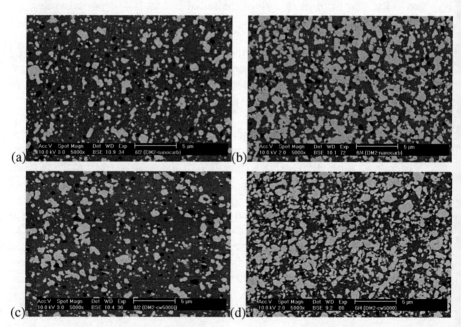

Fig 2. Backscattered electron micrographs revealing the microstructure of 20 (a) and 40 vol % (b) Nanocarb WC powder based and 20 (c) and 40 vol % (d) CW5000 powder based Y-TZP/WC composites. The phases that can be differentiated are WC (white), ZrO_2 (grey) and Al_2O_3 (black).

Since the fracture toughness of Y-TZP materials can be modified by adjusting the Y_2O_3 stabiliser content, the influence of changing the overall yttria content on the fracture toughness was investigated for ZrO_2-WC (60/40) composites. The change in overall yttria content was established by mixing monoclinic and 3 mol % yttria-stabilised ZrO_2 powders in the appropriate amount. The obtained results, given in Table I, clearly reveal that the fracture toughness of the composites increases with decreasing yttria content from 3 to 2 mol %. At an overall yttria content of 1.75 mol % however, the ZrO_2 phase in the samples spontaneously transforms to m-ZrO_2, resulting in macrocracking of the hot pressed material. An excellent fracture toughness of 9.2 MPa $m^{1/2}$ was obtained for the 2Y-TZP-WC (60/40) composite. The hardness of the composites was hardly influenced by a reduction of the yttria content down to 2 mol %, whereas the E-modulus decreased with decreasing yttria content.

Because of the excellent fracture toughness obtained with the 2Y-TZP based composites, this matrix was selected to further investigate the influence of the amount and size of the WC phase. The density, E-modulus, hardness and fracture toughness of 2 mol % Y_2O_3-stabilised ZrO_2-WC composites are summarised in Table II. The hardness of the composites was strongly influenced by the WC content and increased linearly with increasing WC content from 12.28 GPa (HV$_{10}$) for the Y-TZP ceramic up to 16.40 GPa for the 50 vol % WC composite. The hardness of the 40 vol % WC nanopowder based composite is about 2.50 GPa higher than that of the micron-sized powder based composite, whereas the hardness of the 20 vol % ZrO_2-WC composites is comparable.

Beside the high hardness, excellent toughness values of more than 8 MPa $m^{1/2}$ were obtained. For the 2Y-TZP and ZrO_2-WC (80/20) composite, the toughness was as high as 10 MPa $m^{1/2}$. The toughness of the ZrO_2-WC (70/30) and (60/40) composites was comparable, i.e., 9.2 MPa $m^{1/2}$. The toughness of the ZrO_2-WC (50/50) composite was 8.2 MPa $m^{1/2}$. The fracture toughness of the composites with the nanograined WC is higher than that of the composites with micron-sized WC.

Table I. Mechanical properties of Y-TZP/WC composites with 40 vol % nanosized WC starting powder as a function of the overall Y_2O_3 content

Y_2O_3 Mol %	ρ g/cm^3	E GPa	HV$_{10}$ GPa	K_{IC} MPa m$^{1/2}$
3.00	9.94	378	15.13 ± 0.19	6.0 ± 0.2
2.50	9.90	353	15.70 ± 0.20	6.5 ± 0.3
2.00	9.99	322	15.16 ± 0.20	9.2 ± 0.3
1.75	Spontaneous transformation of the composite			

Table II. Mechanical properties of 2Y-TZP/WC composites

WC grade	WC Vol %	ρ g/cm^3	E GPa	HV$_{10}$ GPa	K$_{IC}$ MPa m$^{1/2}$
Nanocarb	0	6.06	203	12.28 ± 0.11	10.1 ± 0.1
Nanocarb	20	8.13	268	13.33 ± 0.06	9.9 ± 0.3
Nanocarb	30	9.03	265	14.01 ± 0.06	9.2 ± 0.2
Nanocarb	40	9.99	322	15.16 ± 0.20	9.2 ± 0.3
Nanocarb	50	10.71	365	16.40 ± 0.44	8.1 ± 0.3
CW5000	20	7.93	259	13.50 ± 0.23	8.7 ± 0.3
CW5000	40	9.95	304	13.56 ± 0.29	7.9 ± 0.2

DISCUSSION

Unlike in other Y-TZP composite systems with TiB$_2$, TiN, TiC or TiCN additions, where the hardness is rather modest (HV$_{10}$ = 12-13 GPa) and hardly influenced by the amount of secondary phase content up to 40 vol % [13], the hardness in the ZrO$_2$-WC system clearly increases with increasing WC content. The high hardness in the ZrO$_2$-WC composites can be explained by the good coherence between WC and the ZrO$_2$ matrix [20] and the chemical compatibility of WC and ZrO$_2$ [21], whereas the Ti-based additives were found to partially dissolve in the ZrO$_2$ matrix [13].

Beside the volume fraction of the WC in the composites, the size of the WC particles strongly influences the hardness of the composites. For the composites with 40 vol % WC, the composite with the micron-sized CW5000 powder had a hardness of 13.56 GPa, whereas the hardness of the composites with the nanosized WC starting powder was significantly higher, i.e. 15.16 GPa, as summarised in Table II.

Additional information on the transformability of the ZrO$_2$ matrix in the optimised 2Y-TZP based composites could be obtained from XRD analysis of polished and fractured ceramic surfaces. The m-ZrO$_2$ fraction measured on polished and fractured materials is summarised in Table III. The reported values were obtained from at least 5 surfaces. The transformability of the zirconia phase, defined as the fraction of the ZrO$_2$ matrix that can be transformed into m-ZrO$_2$ upon fracturing, can be calculated from the m-ZrO$_2$ phase content measured on fractured and polished surfaces. The transformability of the composite is defined as the fraction of t-ZrO$_2$ in the composite that transforms during fracturing, taking into account the presence of the secondary WC phase.

The transformability of the tetragonal zirconia phase decreased from 63 % in the pure 2Y-TZP down to 41 % in the composite with 50 vol % WC, as shown in Table III. Accordingly, the transformability of the composite decreases with increasing WC content. The reduced transformability can be attributed to the increasing WC phase content, reducing the size of the transformation zone around propagating cracks.

Table III. ZrO_2 phase and composite transformability in 2YTZP-WC composites

WC Vol %	$*V_m^{polished}$ (%)	$*V_m^{fractured}$ (%)	ZrO_2 phase transform-ability (%)	$\circ M^{polished}$ (vol %)	$\circ M^{fractured}$ (vol %)	Composite transform-ability (%)
0	8.4 ± 0.6	71 ± 2	63 ± 2	8.4 ± 0.6	71 ± 2	63 ± 2
20	6.5 ± 0.8	67 ± 4	61 ± 1	3.3 ± 0.5	53 ± 3	51 ± 1
30	7.2 ± 0.9	60 ± 1	53 ± 1	3.6 ± 0.5	42 ± 1	38 ± 1
40	8.5 ± 0.4	54 ± 2	46 ± 2	4.2 ± 0.3	33 ± 1	28 ± 1
50	7.9 ± 0.8	49 ± 2	41 ± 2	3.9 ± 0.4	25 ± 1	21 ± 1

* : m-ZrO_2 fraction in the ZrO_2 matrix measured on polished and fractured surfaces
$\circ$: Vol % m-ZrO_2 in the composite, measured on polished and fractured surfaces

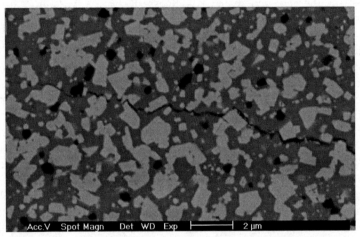

Fig 3. Backscattered electron micrograph revealing crack deflection and crack bridging by the WC grains in a ZrO_2-WC (60/40) composites.

Beside transformation toughening, scanning electron microscopy investigation of the radial crack pattern originating at the corners of Vickers indentations revealed that propagating cracks were deflected by the WC grains, as shown in Fig. 3 for a ZrO_2-WC (60/40) composite. Microcracks and crack branching were not observed at the crack front. In literature however, the major toughening mechanisms in ZrO_2-WC composites were reported to be microcracking, crack bridging, crack deflection and crack branching [16].

The overall toughness of the composites can be described as:

$$K_{Ic} = K_0 + \Delta K_C \qquad (1)$$

with K_0, the toughness of the non-transforming matrix and ΔK_C, the contribution of the different active toughening mechanisms. Based on the observed mechanisms in the 2Y-TZP/WC composites, this can be rewritten as:

$$\Delta K_c = \Delta K_{cT} + \Delta K_{cD} + \Delta K_{cC} \qquad (2)$$

with ΔK_{CT}, the contribution from transformation toughening, ΔK_{CD}, the crack deflection contribution and ΔK_{CC} the contribution of other active toughnening mechanisms

The experimentally measured indentation fracture toughness, K_{IC} (10 kg), for a non-transformable Daiichi HSY-3U powder based Y-TZP ceramic, hot pressed for 1 hour at 1450°C, was 3.5 ± 0.1 MPa m$^{1/2}$. An experimental value which is very close to the reported matrix toughness, $K_0 = 3.3$ MPa m$^{1/2}$ for Y-TZP ceramics [1,22].

From the measured fracture toughness of 10.1 MPa m$^{1/2}$ and a ZrO$_2$ phase transformability of 63 % for the pure 2Y-TZP material, assuming a matrix toughness of 3.5 MPa m$^{1/2}$, the effective contribution of transformation toughening can be calculated from the composite transformability, calculated from the measured ZrO$_2$ phase transformability (see Table III). The results of these calculations are graphically presented in Fig. 4.

From Fig. 4, it is clear that transformation toughening is the primary toughening mechanism in the pure 2Y-TZP, and the composites with 20 and 30 vol % WC. The matrix toughness, transformation toughening and toughening by combined crack deflection and other toughening mechanisms are equally important in the ZrO$_2$-WC (60/40) composites with 40 and 50 vol % WC. The importance of the crack deflection and crack bridging mechanism increases with increasing WC content. The crack deflection model of Faber and Evans [23] predicts a toughness increase of 15 % for ceramic composites with 30 volume % of spherical secondary phase particles, which is severely underestimating the experimentally observations for the ZrO$_2$-WC (70/30) composite, since the toughness increase due to other toughening mechanisms than transformation toughening accounts for about 48 % of the K_0 in the ZrO$_2$-WC (70/30) composites. An important feature of the crack deflection analysis by Faber and Evans [23] is the slightly increasing contribution of crack deflection at volume fractions in excess of 20 vol %, which is in agreement with the experimentally observed increased combined $\Delta K_{CD} + \Delta K_{CC}$ contribution with increasing WC content, as shown in Fig. 4.

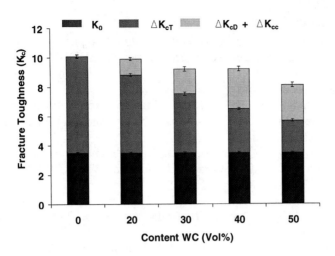

Fig 4. Contribution of the observed toughening mechanisms to the fracture toughness of 2Y-TZP/WC composites. ΔK_{CT} = transformation toughening, ΔK_{CD} = crack deflection, ΔK_0 = toughness of the non-transformable ZrO_2 matrix and ΔK_{CC} = other toughening mechanisms.

CONCLUSIONS

Fully dense ZrO_2-WC composites with excellent toughness > 8 MPa m$^{1/2}$ and hardness > 14.50 GPa were prepared from ZrO_2 and WC nanopowders by means of hot pressing at 1450°C.

The hardness as well as fracture toughness of the composites prepared with WC nanopowders was substantially higher than that of the composites based on micron-sized WC powder.

Composites with optimum fracture toughness were obtained by careful engineering of the ZrO_2 matrix. A maximum toughness was achieved with an overall yttria stabiliser content of 2 mol %, established by mixing pure monoclinic and 3 mol % Y_2O_3 co-precipitated ZrO_2 starting powders.

Experimental observations allowed to evaluate the relative contribution of the matrix toughness, transformation toughening and combined crack deflection and bridging to the fracture toughness of the composites as a function of the WC content. Transformation toughening is important in all investigated composites with a WC content up to 50 vol % and is the primary toughening mechanism in the composites up to 30 vol % WC. The contribution of combined crack deflection and other toughening mechanisms increases with increasing WC content and becomes as important as the ZrO_2 matrix toughness and transformation toughening in the ZrO_2-WC (60/40) composites.

ACKNOWLEDGEMENT
This work was supported by the Brite-Euram program of the Commission of the European Communities under project contract No. BRPR-CT97-0432.

REFERENCES
[1]R.H.J. Hannink, P.M. Kelly and B.C. Muddle, "Transformation Toughening in Zirconia-Containing Ceramics," *Journal of the American Ceramic Society*, **83** [3] 461–87 (2000).
[2]Y. Qi-Ming, T. Jia-Qi and J. Zheng-Guo, "Preparation and Properties of Zirconia-Toughened Mullite Ceramics," *Journal of the American Ceramic Society*, **69** [3] 265-67 (1986).
[3]M. Knutson-Wedel, L.K.L. Falk and T. Ekström, "Characterization of Si3N4 Ceramics Formed with Different Oxide Additives," *Journal of Hard Materials*, **3** [3-4] 435-45 (1992).
[4]R. Telle, S. Meyer, G. Petzow and E.D. Franz, "Sintering Behaviour and Phase Reactions of TiB_2 with ZrO_2 Additives," *Materials Science and Engineering*, **A105/106** 125-29 (1988).
[5]R. Telle and G. Petzow, "Strengthening and Toughening of Boride and Carbide Hard Material Composites," *Materials Science and Engineering*, **A105/106** 97-104 (1988).
[6]T. Watanabe and K. Shoubu, "Mechanical Properties of Hot Pressed TiB_2-ZrO_2 Composites," *Journal of the American Ceramic Society*, **68** [2] C34-C36 (1985).
[7]K. Shoubu, T. Watanabe, J. Drennan, R.H.J. Hannink and M.V. Swain, "Toughening Mechanisms and Microstructures of TiB_2-ZrO_2 Composites"; pp. 1091-99 in *Advances in Ceramics*, volume 24: Science and Technology of Zirconia III, Proceedings of the 4 th International Conference on the Science and Technology of Zirconia (Tokyo, Japan, 1986). Edited by S. Somiya, N. Yamamoto, H. Hanagido, The American Ceramic Society, Columbus, OH, 1986.
[8]V. Gross, J. Haylock and M.V. Swain, "Transformation Toughened Titanium Carbo-Nitride Zirconia Composites"; pp. 555-59 in *Ceramic Developments, Materials Science Forum* Volumes 34-36. Edited by C.C. Sorrell and B. Ben-Nissan. Trans Tech Publications Ltd., Lausanne, Switzerland, 1988.
[9]E. Barbier and F. Thevenot, "Titanium Carbonitride-Zirconia Composites: Formation and Characterisation," *Journal of the European Ceramic Society*, **8** 263-9 (1991).
[10]M. Fukuhara, "Properties of $(Y)ZrO_2$-Al_2O_3 and $(Y)ZrO_2$-Al_2O_3-(Ti or Si)C Composites," *Journal of the American Ceramic Society*, **72** [2] 236-42 (1989).
[11]S.E. Dougherty, T.G. Nieh, J. Wadsworth and Y. Akimune, "Mechanical Properties of a 20 vol % SiC Whisker-Reinforced Yttria-Stabilised Tetragonal

Zirconia Composite at Elevated Temperatures," *Journal of Materials Research*, 10 [1] 113-18 (1995).

[12]Z. Pędzich, K. Haberko, J. Babiarz and M. Faryna, "The TZP-Chromium Oxide and Chromium Carbide Composites," *Journal of the European Ceramic Society*, 5 1939-43 (1998).

[13]J. Vleugels and O. Van Der Biest, "Development and Characterisation of Y-TZP Composites with TiB₂, TiN, TiC and TiCN," *Journal of the American Ceramic Society*, 82 [10] 2717-20 (1999).

[14]Z. Pędzich, K. Haberko, J. Piekarczyk, M. Faryna and L. Litynska, "Zirconia Matrix-Tungsten Carbide Particulate Composites Manufactured by Hot Pressing Technique," *Materials Letters*, 36 70-75 (1998).

[15]K. Haberko, Z. Pędzich, G. Róg, M. Bucko and M. Faryna, "The TZP Matrix - WC Particulate Composites," *European Journal of Solid State Inorganic Chemistry*, 32 [7-8] 593-601 (1995).

[16]Z Pędzich and Haberko K., "Toughening Mechanisms in the TZP-WC Particulate Composites," *Key Engineering Materials*, 132-136, 2076-79 (1997).

[17]G.R. Anstis, P. Chantikul, B.R. Lawn and D.B. Marshall, "A Critical Evaluation of Indentation Techniques for Measuring Fracture Toughness: I, Direct Crack Measurements," *Journal of the American Ceramic Society*, 64 [9] 533-38 (1981).

[18]ASTM E 1876-99, American Society of Testing and Materials (2000).

[19]Lin J.D. and Duh J.G., "The Use of X-ray Line Profile Analysis in the Tetragonal to Monoclinic Phase Transformation of Ball Milled, As-Sintered and Thermally Aged Zirconia Powders," *Journal of Materials Science*, 32 4901-8 (1997).

[20]M. Faryna, K. Haberko, J.A. Kozubowski and Z. Pędzich, "Interphase Boundary in the TZP-WC Composites," pp. 745-46 in *Electron Microscopy 1998*, volume II. Proceedings of ICEM14 (Cancun, Mexico, 1998). Edited by H.A. Calderón Benavides and M. J. Yacamán. Institute of Physics Publishing, Philadelphia, 1998.

[21]K. Haberko, Z. Pędzich, J. Piekarczyk and G. Rog, "Zirconia-Tungsten Carbide Particulate Composites Part 1: Manufacturing and Physical Properties", Fourth Euro Ceramics-Vol 4, pp. 29-36, 1995, edited by A. Bellosi, Italy, Gruppo Editoriale Faenza Editrice S.p.a.

[22]M. V. Swain and L. R. F. Rose, "Strength Limitations of Transformation-Toughened Zirconia Alloys," *Journal of the American Ceramic Society*, 69 [7] 511–18 (1986).

[23]K. T. Faber and A. G. Evans, "Crack Deflection Processes - I Theory," *Acta Metallurgica*, 31 [4] 565-76 (1983).

PRESSURELESS SINTERING OF Al$_2$O$_3$-TiC POWDERS PRODUCED FROM CARBON COATED PRECURSORS

Hisashi Kaga, Kevin B. Newman, and Rasit Koc
Department of Mechanical Engineering and Energy Processes,
Southern Illinois University at Carbondale
Carbondale, IL 62901-6603, USA

ABSTRACT

The sintering behavior of submicron/nanosize Al$_2$O$_3$-TiC composite powders was investigated. The Al$_2$O$_3$-TiC composite powders were synthesized from carbon coated TiO$_2$/Al mixture[1]. The synthesized powders with a surface area of 22m^2/g were obtained. Using these powders, pressureless sintering of MgO doped Al$_2$O$_3$-20wt.%TiC was accomplished at 1650°C for 2 hrs. in flowing argon gas and showed that densification of 98% TD (Theoretical Density) and hardness in excess of 17GPa were achieved. Results showed that synthesized powder from carbon coated of Al/TiO$_2$ mixture could be presureless sintered. The hardness of these composites were comparable with those of hot-pressed Al$_2$O$_3$-TiC composites.

INTRODUCTION

Composites of Al$_2$O$_3$-TiC consist of dispersed TiC in an Al$_2$O$_3$ matrix and have high temperature strength, high wear resistance, good corrosion resistance, and other special physical properties. It is used in the machining as well as in the electronic and computer industry[2-7]. The major advantage of adding TiC to Al$_2$O$_3$ is that the carbide limits the grain growth of the Al$_2$O$_3$ matrix and results in a higher strength and hardness[4]. To improve the sintering behavior of the composites, fine particle size is required. Smaller particles enhance the capillary forces, despite a higher friction between the particles. Similarly, the densification rate is increased when using fine powders. In addition, the starting powders must be pure and single phase, have a narrow size distribution, be spherical in shape, and be free of agglomerations. Failure in any of these criteria lessens the quality of the final sintered piece[8]. The first step in this research consisted of the production of nanosize Al$_2$O$_3$-TiC powders using the carbon coated precursor method[1]. The second step consisted of sintering and characterizing the sintered pellets.

EXPERIMENTAL PROCEDURE

Powder Preparation:

The nanosize Al_2O_3-TiC composite powders are produced utilizing the carbon coating process which consists of two steps[1]. In the first step, mixture of TiO_2 powder (P-25, Degussa Corp., Ridgefield Park, NJ) and Al powder (41000, Alfa Aesar, Ward Hill, MA) was coated with carbon by decomposing a hydrocarbon gas (C_3H_6). In the second step, Al_2O_3-TiC is formed by promoting the aluminothermic reaction of the carbon coated TiO_2/Al mixture at $1200°C$ under inert gas a tube furnace (Model CTF 17/75/300, Carbolite, Sheffield, UK)[8].

The Al_2O_3-TiC from carbon coated TiO_2/Al mixture was characterized using BET surface area analyzer (Micromeritics, Gemini 2360), a X-ray diffraction (Rigaku, Tokyo, Japan) with CuK_α radiation, Transmission Electron Microscopy (TEM)(Hitachi, Model FA 7100, Tokyo, Japan), Scanning Electron Microscopy (SEM) (Hitachi, S570, Tokyo, Japan), and Energy Dispersive X-Ray (EDS) (Noran, Model Voyager, Middleton, WI).

Sintering of Al_2O_3-TiC Powders:

The composites were sintered of Al_2O_3-TiC from carbon coated precursor. The synthesized powders (with a surface area of about $22m^2$/g) were mixed with Al_2O_3 powder (Degussa AG, Teterboro, NJ) to make the ratios of 20, 30, and 46.9wt.% TiC. MgO was also added to some of the Al_2O_3-TiC mixtures in order to promote the sintering of the Al_2O_3 portion of the composite. All mixing of synthesized powders was performed by in a ball mill (Model 8000, Spex, Meutuchen, NJ) with WC media and ethyl alcohol. After drying, the powders were mixed with a PVA binder to achieve better binding characteristics. The pellets were produced in a hardened steel die (i.d. = 13mm) using single action, uni-axial pressing at a pressure of 440MPa. The green density was measured at a range of 53% to 56% of TD. Then sintering was performed at $1650°C$ for 2 hours under

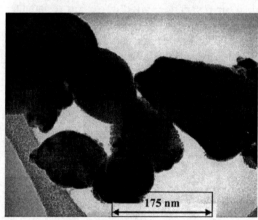

175 nm

Fig. 1: Typical Al_2O_3-TiC powders

flowing argon and kept there for 2 hours. Heating rate of 2°C/min up to 800°C and then 10°C/min was used. The bulk densities of the sintered pellets were determined by applying the liquid displacement technique. The pellets were polished and microstructure was examined with a SEM. Vickers hardness was determined with Shimadzu HSV 20 hardness tester.

RESULTS AND DISCUSSION

A very uniform coating was the result of the carbon coated method. The BET surface area of the uncoated TiO_2/Al mixture was measured to be $35m^2/g$. After being coated with 9.3wt.% carbon, the surface area was determined to be $31m^2/g$. These results show that a uniform carbon coating on both TiO_2 and Al particles can be obtained by using a hydrocarbon gas. The coated precursor was synthesized at 1200°C for 2 hrs. under flowing argon with a heating rate of 4°C/min. The synthesized powders with a surface area of $22m^2/g$ were obtained.

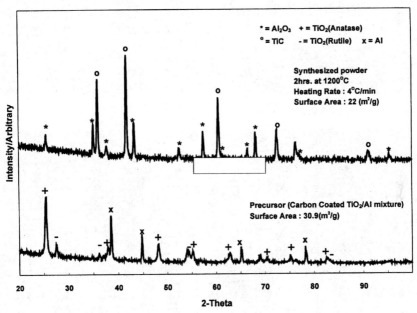

Fig.2 XRD Patterns of synthesized powder and precursors

This reaction inhibits strong exothermic and short reaction time which is usually observed during the traditional method of mixing Al, TiO_2, and Carbon black. The result of our method was nanosize Al_2O_3-TiC powder with a uniform size and loose agglomeration. Furthermore, iron impurities contained in the carbon black are eliminated because the carbon source is hydrocarbon gas. The typical Al_2O_3-TiC composite powders produced from carbon coated method used in this study can be seen in the Fig. 1. The high purity and surface area of the synthesis powder is displayed in the XRD pattern with its precursor. As it can be seen from the XRD patterns (Fig.2), all starting powders, which are TiO_2, Al, and C, converted to Al_2O_3 and TiC at 1200°C.

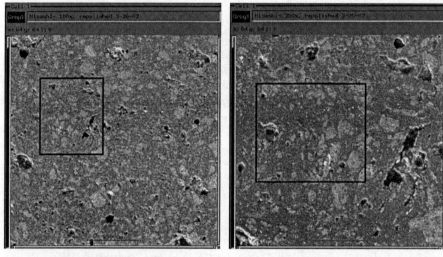

Fig. 3 SEM micrograph (x100)*   Fig. 4 SEM micrograph (x250)*

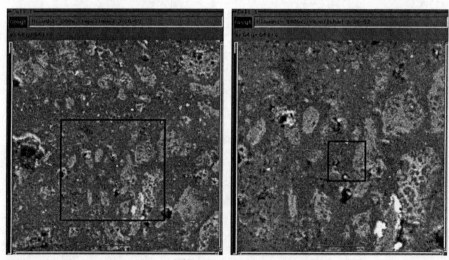

Fig. 5 SEM micrograph (x500)*   Fig. 6 SEM micrograph (x1000)*
*All SEM figures are of Al_2O_3-20wt.%TiC with 0.5wt.%MgO of Al_2O_3

Sintering of the compact was performed in a box furnace (CM Furnace, Bloomfield, NJ) under flowing argon gas at 1650°C for 2 hrs. Specimens were embedded in alumina powder (Degussa AG, Teterboro, NJ) to help prevent reaction with the environment during sintering. The relative density of the sintered and green pellets is shown in Table I.

Table I: Green and sintered density with their hardness

Specimen	Green Density (%TD)	Final Density (%TD)	Hardness (GPa)
Al_2O_3-20wt.%TiC*	53.6	98.1	17.2
Al_2O_3-30wt.%TiC*	54.1	93.1	19.8
Al_2O_3-46.9wt.%TiC*	56.6	89.3	21.5
Al_2O_3-20wt.%TiC	52.5	87.0	16.4
Al_2O_3-30wt.%TiC	53.6	89.6	19.0
Al_2O_3-46.9wt.%TiC	55.6	75.3	20.8

*0.5wt.%MgO of Al_2O_3

As shown in Table I, the density increased with the addition of MgO. The increase was large with 20wt.%TiC composite, and more than 98% of theoretical density was obtained, which is comparable with the result by hot-pressing even though sintering temperature and heating rate are relatively low[6]. In addition, Vickers hardness value of 17.2GPa is also comparable with commercially available composites. SEM photographs of the polished surface with the highest density of the specimen are shown in Fig. 3 to Fig. 5. The white grains are TiC ranging in size 5 – 10 μm with some agglomerations. This result shows that the grain growth was found in the sintered bodies due to low heating rate and long duration time during the pressureless sintering.

CONCLUSION

The Al_2O_3-TiC powders from carbon coated precursor show that they are single phase and of high purity and surface area. The MgO doped Al_2O_3-20wt.%TiC resulted in a 98% TD after pressureless sintering with relatively low sintering temperature and heating rate. However, during the sintering, the grain growth of TiC was obtained. It is postulate that relatively low heating rate and long sintering time (2°C/min and 10°C/min) will lead TiC to growth. Usually, temperatures in excess of 1850°C and heating rate of 40 - 50°C/min are required to densify the oxide-carbide composites[6]. The sintering behavior of these powders during hot pressing will be analyzed in further studies to avoid grain growth during pressureless sintering, as it has been shown that hot-pressing leads to a decrease in grain size[9].

ACKNOWLEDGEMENT
This research is sponsored by the U.S. Department of Energy, Assistant Secretary for Energy Efficiency and Renewable Energy, Office of Industrial Technologies, Advanced Industrial Materials Program, under contract DE-AC05-96OR22464 with Lockheed Martin Energy Research Corporation/UT-Batelle.

REFERNECES
[1] R. Koc and G. Glatzmaier, US Patent No. 5,417,952, (1995)
[2] A.G. King, *Am. Ceram. Soc. Bull.* 43, (1965), 395
[3] D. Bordui, *Am. Ceram. Soc. Bull.* 67, (1988), 998
[4] R. P. Wahi and B. Ilschner, *J. Mater. Sci.* 15 (1980) 875
[5] R. A. Culter, A. V. Virkar and J. B. Holt, *Cerem. Eng. Sci. Proc.,* 6 (1985) 715
[6] R. A. Culter *et al, Mater. Sci. Eng. A* 105/106 (1988) 183
[7] S. J. Burden, *Am. Ceram. Soc. Bull.* 67 (1988) 1003
[8] H. Kaga, K. B. Newman, and R. Koc, *ACerS 104th Annual Meeting, Paper No. AMC.4-H-05-2002,* St. Louis, MO, 4/28-5/1, 2002.
[9] M.N. Rahaman, "Ceramic Processing and Sintering", Marcel Dekker Inc., New York, 1995

Process Modeling

FINITE ELEMENT SIMULATION OF COMPACTION AND SINTERING OF TILES HAVING TWO LAYERS OF POWDERS

Manabu Umeda
INAX Corporation
3-77 Minato-cho, Tokoname,
Aichi 479-8588, Japan

Ken-ichiro Mori and Morito Murakami
Toyohashi University of Technology
1-1 Tempaku-cho, Toyohashi,
Aichi 441-8580, Japan

ABSTRACT

The densification behaviors in compaction and sintering processes of tiles having two layers of powders were simulated by the viscoplastic finite element method. The flow stress of the powder and the shrinkage strain-rate during the sintering required as material constants for the simulation were measured from a simple compression test with constant load. The flow stress with strain-rate and temperature sensitivities and the history of shrinkage strain-rate were determined. Using these material constants, the accuracy of the calculated results was improved.

INTRODUCTION

Tiles used as building materials are manufactured by compacting ceramic powders in the dies and then sintering the compacts at high temperature. The sintered product is bent and distorted by a non-uniform density distribution of the compact due to the forming of grooves of the upper surface of tiles. Because the tiles are composed of colored upper layer and uncolored lower layer, the sintered products also undergo non-uniform shrinkage during the sintering due to the difference between shrinkages of the two layers. The powder of the uncolored lower layer is cheaper than that of the colored upper layer, and thus the cost is reduced by the use of the cheap powder in the lower layer. It is desirable in ceramic industry to develop a method for simulating the densification behavior in sintering of ceramic compacts. The information obtained from the numerical simulation is effective in designing the sintering process.

In the field of metal forming, the finite element simulation is generally employed for the design of processes. Mori et al.[1] have proposed a viscoplastic finite element method for simulating non-uniform shrinkage in sintering of ceramic powder compact. In the finite element method, viscoplastic deformation induced by difference of the shrinkage strain is calculated, and the effects of the density distribution, the dead weight and the different powders are taken into consideration. Using calculated results by the finite element simulation, the occurrence of fracture caused by the non-uniform shrinkage in the sintering of ceramic compacts has been predicted[2]. In addition, the finite element simulation has been applied to net shape forming for obtaining desired shapes of the sintered products[1]. On the other hand, Riedel et al.[3] and Shinagawa[4] have introduced the sintering stress induced by the surface tension into the formulation of finite element methods for sintering processes. The method using the shrinkage strain is more useful than that using the sintering stress because of easy measurement of shrinkage strain.

In the finite element simulation, the flow stress, the shrinkage strain-rate, etc. are required as material constants. Since the temperature in sintering is very high, it is not easy to measure the material constants. Mori et al.[1] have determined the flow stress from sintering of cantilevered sheet compacts using calculated results by the finite element simulation. The ratio of flow stress for two powders has been obtained from a sintering test of a cylindrical compact inserted into a rigged compact. In the conventional simulation, the exponent of strain-rate sensitivity in the flow stress during sintering is assumed to be 1, namely the Newtonian fluid, like liquid because of the simplification. The exponent in the sintering without melting of the powder may be smaller than 1.

In the present study, a simple compression test with constant load is presented to measure the shrinkage strain-rate and the flow stress of powders during the sintering. Using the measured material constants, non-uniform shrinkage behavior in sintering of tiles is simulated by the viscoplastic finite element method.

FINITE ELEMENT METHOD FOR SINTERING SHRINKAGE
In the sintering of powder compacts, non-uniform shrinkage is caused by the density distribution, the dead weight, the external pressure, etc. To simulate the non-uniform shrinkage in the sintering of tiles, the viscoplastic finite element method formulated by Mori et al.[1] is employed. In this method, the total strain-rates in sintering are assumed to be composed of two parts

$$\{\dot{\varepsilon}\}=\{\dot{\varepsilon}^s\}+\{\dot{\varepsilon}^p\}, \tag{1}$$

where is the matrix $\{\dot{\varepsilon}^s\}$ is the vector of shrinkage strain-rate representing the normal strain-rate.

The total strain-rates within the element are expressed by

$$\{\dot{\varepsilon}\}=[B]\ \{v_e\}, \tag{2}$$

where [B] is the matrix correlating the strain-rate with the nodal velocity.

The ceramic compact in sintering is assumed to be a viscoplastic porous material. The constitutive equations for the viscoplastic porous materials[5] are expressed by

$$\{\sigma\}=[D]\{\dot{\varepsilon}^P\},\tag{3}$$

where [D] is the matrix correlating the stress with the plastic strain-rate. By substituting Eqns. (1) and (2) into Eqn. (3), the stresses in sintering are given by

$$\{\sigma\}=[D][B]\{v_e\}-[D]\{\dot{\varepsilon}^s\}.\tag{4}$$

From the principle of virtual work, the nodal forces for each element are obtained

$$\{P\}=\int_{V_e}[B]^T\{\sigma\}dV+\int_{V_e}g\rho\gamma\{N\}dV,\tag{5}$$

where g is the gravity acceleration, ρ is the relative density, γ is the density and $\{N\}$ is the shape function of the element.

By substituting Eqn. (4) into Eqn. (5), we have

$$\{P\}=\left[\int_{V_e}[B]^T[D][B]dV\right]\{v_e\}-\int_{V_e}[B]^T[D]\{\dot{\varepsilon}^s\}dV+\int_e g\rho\gamma\{N\}dV.\tag{6}$$

The three terms on the right hand side of Eqn. (6) represent the effects of viscoplastic deformation, the sintering shrinkage and the dead weight, respectively. The nodal forces of the surrounding elements at each nodal points are equilibrated

$$\sum^{element}P_i=\begin{cases}0\ (\text{ in material })\\F_i(\text{ on surface })\end{cases},\tag{7}$$

where F_i is the external force obtained from the external pressure and from the friction between the base and compact.

The shrinkage strain-rate $\dot{\varepsilon}^s$ in Eqn. (6) is the material constant determined from the experiment. The present method acquires a solution for a given distribution of the shrinkage strain-rate. The shrinkage behavior due to the rise of temperature in sintering is treated as the volumetric strain in the formulation.

MEASUREMENT OF MATERIAL CONSTANTS
Determination of Flow Stress

In the finite element simulation, the flow stress and the shrinkage strain-rate in the sintering are required as material constant. These material constants were measured from a simple compression test with constant load as shown in Fig. 1. In this test, a compact is sintered under setting the weight. The compression test without load is a mere sintering, whereas viscoplastic deformation is induced by increasing the load. In the measurement, the thermal expansion of detector is canceled using another detector with the standard specimen.

From the yield criterion for porous materials[5], the flow stress for the simple compression test is expressed by

$$\overline{\sigma} = 1/\rho^k \left\{ \sigma_z^2 + \left(\sigma_z^2 / 9f^2 \right) \right\}^{0.5},$$

$$f = 1/2.5(1-\rho)^{0.5}$$

$$(8)$$

where ρ is the relative density, k is the material constant determined by powder compaction.

The equivalent strain-rate for the simple compression test is expressed by

$$\dot{\overline{\varepsilon}} = \rho^{k-1} \left[4/9(\dot{\varepsilon}_x - \dot{\varepsilon}_z)^2 + \left\{ f(2\dot{\varepsilon}_x + \dot{\varepsilon}_z) \right\}^2 \right]^{0.5}. \qquad (9)$$

The flow stress curve is determined by Eqns. (8) and (9).

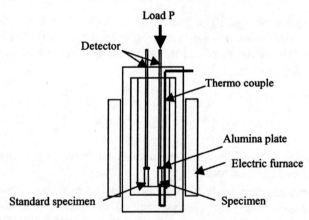

Fig. 1. Simple compression test with constant load for measurement of flow stress and shrinkage strain-rate in sintering.

Testing Conditions

The tiles are composed of colored upper layer and uncolored lower layer, and the ceramic powders of the two layers are mixtures of kaolin, feldspar, sericite, etc. Each powder was compacted into a block of 5x5x20mm with a uniform distribution, and the relative densities ρ_0 are 0.66, 0.7, 0.73, 0.75, 0.78 and 0.80. The compact was sintered under the loads P=49, 98, 196 and 294mN, and the buckling occurred above 294mN. The shrinkage strain-rate in sintering was measured under P=0.98mN to prevent the disturbance due to the oscillation of displacement. The two orthogonal strains are assumed to be proportional to the axial strain.

Histories of Shrinkage Strain-rate

The histories of displacement in the sintering measured by the simple compression test for P=0.98mN are illustrated in Fig. 2. The compact slightly expands in the early stage of sintering because of the dehydration from the kaolin. The shrinkage begins from about 800 °C due to the melting of the feldspar and stops around 1250 °C. The shrinkage of the lower layer is larger than that of the upper layer.

The histories of axial shrinkage strain-rate in sintering are given in Fig. 3. In the finite element simulation, the shrinkage strain-rate is obtained by the linear interpolation for the relative density.

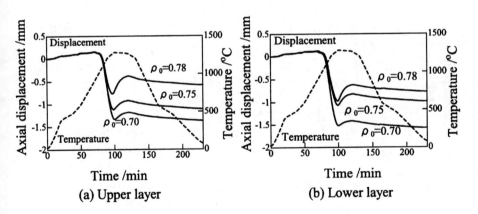

(a) Upper layer   (b) Lower layer

Fig. 2. Histories of displacement in sintering measured by the simple compression test for P=0.98mN.

Flow Stress Curve

The histories of axial strain-rate for viscoplastic deformation for different loads are shown in Fig. 4. The axial strain-rate for viscoplastic deformation is obtained by subtracting the displacement of P=0.98mN from the measured one. The axial strain-rate for viscoplastic deformation becomes large above 800°C, and thus an approximate curve of flow stress is determined in the range above 800°C.

The flow stress is approximated as the following function of equivalent strain-rate and temperature

$$\overline{\sigma} = F \exp(C/T)\dot{\overline{\varepsilon}}^m , \tag{10}$$

where F and C are material constants, m is the exponent of strain-rate sensitivity and T is the absolute temperature.

The measured flow stress is approximated into Eqn.(10) by the multiple regression statistical analysis. The determined material constants and approximate curve are shown in Table 1 and Fig. 5, respectively. The flow stresses for the heating and cooling are approximated as different curves. It is found that the exponent of strain-rate sensitivity, m, is smaller than 1. This means that the conventional simulation using m=1 yields a large error of the calculated results.

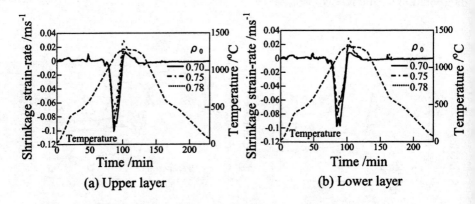

(a) Upper layer   (b) Lower layer

Fig. 3. Histories of axial shrinkage strain-rate in sintering measured by the simple compression test for P=0.98mN.

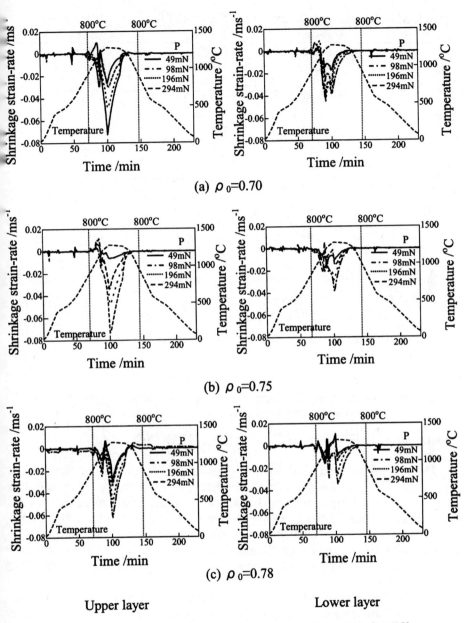

(a) $\rho_0 = 0.70$

(b) $\rho_0 = 0.75$

(c) $\rho_0 = 0.78$

Upper layer   Lower layer

Fig. 4. Histories of axial strain-rate for viscoplastic deformation for different loads in sintering measured by the simple compression test.

Table 1. Material constants determined in Eq. (10) of flow stress in sintering.

Layer	Constants	Heating	Cooling
Upper	F /MPa	0.0013	28.847
	C /K	10678	1044
	m	0.459	0.667
Lower	F /MPa	0.00047	33.182
	C /K	18490	243
	m	0.655	0.606

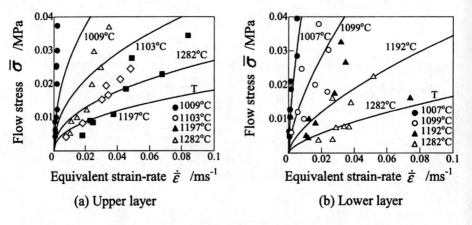

(a) Upper layer   (b) Lower layer

Fig. 5. Measured flow stress in sintering and approximate curve expressed by function of equivalent strain-rate and temperature.

SIMULATION OF COMPACTION AND SINTERING
Compaction

The compaction and sintering processes of two layer circular tiles with a groove shown in Fig. 6 are simulated. The two layer powders are compacted with upper grooved and lower flat punches and a container. The compact has a non-uniform distribution of density due to the press with the grooved punch. Material constants used for the simulation of compaction were measured from a closed die compaction process. The initial relative densities for the upper and lower layers are 0.50 and 0.41, respectively. The powders are compacted up to the height of the compact of 10mm.

The element mesh for the compact with a circular groove and the depth of groove h=2.5mm is shown in Fig. 7. To present severely distorted elements round the corner of the groove, the remeshing is performed in each step of deformation.

The distribution of the calculated relative density for the compact with a circular groove is illustrated in Fig. 8. The relative density is the highest around the corner of groove. As the depth of the groove increases, the relative density becomes large.

The density distribution in compact was measured from the X-ray computed tomography[6]. The relationship between the ratio of X-ray transparency and the relative density beforehand measured for compacts with a uniform distribution is shown in Fig. 9.

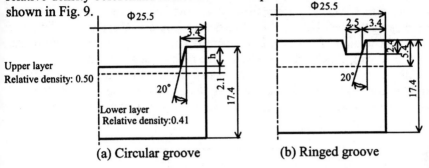

(a) Circular groove   (b) Ringed groove

Fig. 6. Compaction of circular tiles having two powder layers.

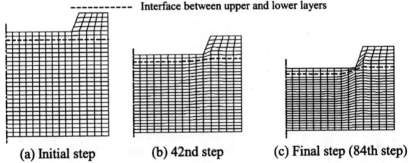

(a) Initial step   (b) 42nd step   (c) Final step (84th step)

Fig. 7. Element mesh in compaction of circular tile with circular groove for h=2.5m

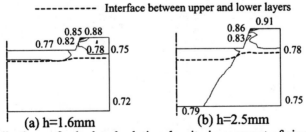

(a) h=1.6mm   (b) h=2.5mm

Fig. 8. Distributions of calculated relative density in compact of circular tile with circular groove.

The distribution of the calculated relative density in the compact with a ringed groove is compared with the experimental one in Fig. 10. The experimental distribution is obtained using Fig. 9. The density measured from the X-ray computed tomography includes error near the surface. The calculated density distribution agrees with the experimental one.

Sintering

Non-uniform shrinkage in sintering of circular compacts with a groove is simulated using the distribution of relative density obtained from the simulation of compaction. The time of one step in the simulation of sintering consisting of the heating and cooling above 800°C is 60s and the number of steps is 84.

The distortion of the element mesh in sintering is shown in Fig. 11. Because the height of the sintered product is not small in comparison with the diameter, the curvature of the sintered product is comparatively small, whereas the resultant stress is not small.

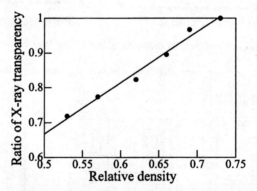

Fig. 9. Relationship between ratio of X-ray transparency and relative density in X-ray computed tomography.

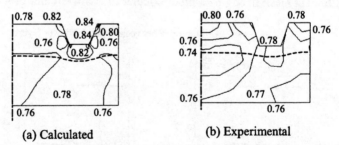

(a) Calculated   (b) Experimental

Fig. 10. Comparison between calculated and experimental density distributions in compact of circular tile with ringed groove.

The distribution of the calculated relative density in the sintered product is illustrated in Fig. 12. The non-uniformity of the distribution around the corner of groove for h=2.5mm is larger than that for h=1.6mm.

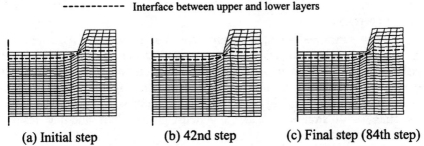

(a) Initial step   (b) 42nd step   (c) Final step (84th step)

Fig. 11. Distortions of element mesh in sintering of circular tile with circular groove for h=2.5mm.

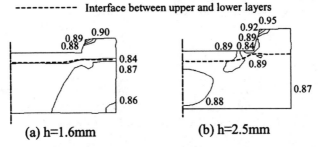

(a) h=1.6mm   (b) h=2.5mm

Fig. 12. Distributions of calculated relative density in circular tile with circular groove after sintering.

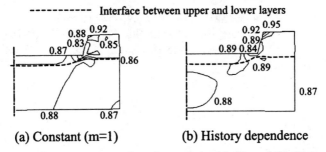

(a) Constant (m=1)   (b) History dependence

Fig. 13. Effects of the exponent of strain-rate sensitivity and the history of shrinkage strain-rate on the distribution of relative density in the sintered product of circular tile with circular groove for h=2.5mm.

The effects of the exponent of strain-rate sensitivity and the history of shrinkage strain-rate on the distribution of relative density in the sintered product are examined on the distribution of the calculated relative density in Fig. 13. The accuracy of the calculated results is improved by the inclusion of these effects.

CONCLUSIONS

The flow stress and shrinkage strain-rate in sintering were measured from a simple compression test with constant load. Using the measured material constants, the densification behavior in compaction and sintering processes of tiles having two layers of powders was simulated by the viscoplastic finite element method. The accuracy of the calculated results was improved by using the flow stress with strain-rate and temperature sensitivities and the history of shrinkage strain-rate.

ACKNOWLEDGMENT

This work was supported in part by the Japanese Ministry of Education, Culture, Sports, Science and Technology with Grant-in-Aid for Scientific Research.

REFERENCES

[1] K. Mori, "Numerical Methods in Industrial Forming Processes NUMFORM'92"; pp. 69-78, Edited by J.- L. Chenot et al., Balkema, 1992.

[2] K. Mori, K. Osakada, and M. Miyazaki, Int. J. Mach. Tools Manufact. 37[9] 1327-36 (1997).

[3] H. Riedel, and D.-Z. Sun, "Numerical Methods in Industrial Forming Processes NUMFORM'92"; pp. 883-86, Edited by J.- L. Chenot et al., Balkema, 1992.

[4] K. Shinagawa, Trans. JSME, 62[593], 246-52 (1996).

[5] S. Shima, and M. Oyane, Int. J. Mech. Sci. 18[2], 285-91 (1976).

[6] Y. Ikeda, Y.Mizuta, H. Tobita and H. Usami, "Proc. 95-YOKOHAMA-IFTC", III; pp. 29-36 (1995).

COMPOSITE BEARING BASED ON "METAL – VISCOUS – ELASTIC MATERIAL (POLYMER) – SOFT METAL – CERAMICS" COMPOSITION

Maksim V. Kireitseu, L.V. Yerakhavets, M.A. Belotserkovski and V.L. Basenuk
Institute of Machine Reliability
National Academy of Sciences of Belarus
Lesnoe 19 – 62, Minsk 223052, Belarus

ABSTRACT
In the paper, composite bearing "alumina – aluminum or their alloy - damping viscous-elastic material - steel substrate" has been discovered. Better understanding of fundamentals of thermal spraying and micro arc oxidizing processes has found semi-empirical relations to be used in the bearing design. It provides desirable mechanical properties and performance.

An analysis of fatigue behavior of the bearing revealed an effect of structure and porosity of alumina on degradation and overall strength of the composition. Of interest is very close correlation between developed rheological model and mechanical behavior of the bearing under indentation. An experimental research of load rating showed a good agreement with rheological model to be applied in analysis of the composite bearing mechanics.

INTRODUCTION
Powder metallurgy of aluminum constructions is developed rapidly. Nowadays composite bearings based on aluminum or its alloys are produced in a wide range [1]. In engineering practice, a lot of composite sliding bearings [2, 3] are have to know comprise of soft base (aluminium, polymer etc.), on which one a layer of friction resistant material like alumina is formed. In some cases, between ceramic layer and base substrate some extra layers are used to form by casting or welding processes. To improve process of friction, top layer of ceramics is impregnated with a solid lubricant. Usually, such multilayered structure has low adhesion between adjacent layers that effect on ultimate shearing stressing, cracks initiation and micro flaws propagation. In practice, hard ceramic coating placed on soft base cannot be overloaded due to significant deformation of ductile soft substrate under applied load. In result all composite layer could be shattered off.

Investigated process is expected to reveal semi empirical relations to calculate thickness of polymer and aluminum layers in the bearing.

OBJECTIVE OF RESEARCH

An objective of the work is to reveal an approach to development of composite bearing "alumina – aluminum – viscous-elastic material (polymer) - steel substrate" through better understanding of fundamentals of thermal flame spraying process and micro arc oxidizing technology. Adequacy of revealed equations has to be checked in experimental researches and practical application.

IDEA: OVERVIEW AND DEVELOPMENT

Over viewing the developed bearing construction consists of sequentially arranged steel substrate, viscous-elastic material (polymer) layer, aluminium or its alloy layer and oxide ceramic layer. Both aluminium and viscous-elastic layers were produced by thermal flame spraying process because by this processing the layers reshape the surface profile reaching strong adhesion (data see below). At optimal regimes, the process does not overheat sprayed materials and substrate. The viscous-elastic material was selected from the polymer group including polyamide, polyvinylchloride, polyethylene, polyethylene(terephthalate). The steel base has ledges and cavities arranged in preferably the quincunx order. The depth of the profile ranges between 1.2-1.8 thicknesses of polymer layer. The layer of aluminium or its alloy was produced with complementary surface. Thickness of Al layer is found to be effective calculating by the following equation (1) detailed in [8, 9]:

$$\delta A = (0.6 \div 0.8) \cdot K \cdot \sqrt[3]{\frac{C_{yd}}{E_a}} \tag{1}$$

Where δA is the minimum depth of aluminium or its alloy, mm. K is a semi-empirical value that depends on the form of processed surface of the bearing. For example, cylindrical sliding bearing has K equal to R, where R is the radius of the working surface of the cylindrical sliding bearing. C_{yd} is the specific contact rigidity of a polymer that is relation of Young's modulus vs. surface area. Ea is Young modulus of aluminium or its alloy.

It has been revealed that thickness of the polymer layer depends on technological regimes applied at deposition of the aluminium layer. Large fragments of Al particles and its temperature should be deposited on thicker layer of polymer to eliminate its over heating, welding and destruction due to thermal impact produced by hot Al particles dropping on polymer surface. The semi empirical relations were revealed through a set of experimental tests conducted before [8,9].

Manufacturing technology of the composite sliding bearing includes step-by-step deposition on the steel base the polymer layer and the layer of aluminium or its alloy produced by thermal flame spraying technology, and then top oxide ceamic layer produced by micro arc oxidizing. To prevent destruction and degradation of mechanical properties of polymer layer in result of impact of heated Al particles the equation has been developed through better understanding of fundamentals and processes of the applied technology.

In this case, thickness of the polymer layer depends on technological regimes to be applied while deposition of the next aluminum layer. Usually, greater size of Al particles have higher temperature should be deposited on thicker polymer layer to prevent localized destruction and degradation of mechanical properties in the polymer in result of heat impact of Al particles. This effect has been revealed in result of the following experimental and theoretical procedure.

In the case when heated Al particle of approximately round shape distributes a heat in a constant mode in the definite time an increase of temperature in a contact point of Al particle with the polymer layer can be described with an increase of temperature in a point of semi-infinite body.

The process of heat distribution can be approximated to a process of heat distribution at a surface of semi-infinite body if the heat during the time

$$\tau_0 = 0.25\, a\kappa \tag{2}$$

distributes at the surface of the body only, and then it distributes in both directions at the surface and in the depth of the body. In the equation (2), a is the thermal conductivity of the polymer layer. κ is coefficient of concentration of heat flow). If it is stated two conditions that (a) the relatively isotropic body does not significantly change parameters of heat flow; and (b) in initial time, the temperature of the body T_0 stills constant in all the body and equals to zero, then the temperature of sprayed Al drop can be shown as the following equation (3):

$$T = \frac{2q}{c\rho\left(4\pi a\tau\right)^{3/2}} e^{-\frac{R^2}{4a\tau}} \tag{3}$$

Reducing a volume of heat transfer of heated drop with an air, the decrease of temperature in rounded space with radius R describes the following coefficient

$$e^{-\frac{R^2}{4a\tau}} \tag{4}$$

Whereas coefficient is found from

$$\frac{q}{c\rho(4\pi a\tau)^{3/2}} \quad (5)$$

and it reveals a change of temperature in the point with radius R = 0 during the time (τ).

Transforming equation (3) by the new elements, it gives the following:

$$\Delta T = \frac{2q}{c\rho} \frac{e^{-\frac{R^2}{4a(\tau+\tau_o)}} e^{-\frac{z^2}{4a\tau}}}{4\pi a(\tau+\tau_o)\sqrt{4\pi a\tau}} \quad (6)$$

Where R is the radius of deformed drop of aluminum; c is heat ρ is density of polymer; q is heat of a drop; τ is time used a drop to reach the polymer layer or substrate; τ_o is initial time when a heat from drop distributes in the depth of polymer layer.

Modifying equation (6) to the base of heat of a drop, it is transformed as follows:

$$q = \frac{c\rho\Delta T 4\pi a(\tau+\tau_o)2\sqrt{\pi a\tau}}{2e^{-\frac{R^2}{4a(\tau+\tau_o)}} e^{-\frac{z^2}{4a\tau}}} \quad (7)$$

The depth of distribution of heat flow from the heated Al drop δ can be described with equation (3) (here the depth δ is measured along the vertical axis Z and it follows that δ=Z):

$$\delta = \left[4a\tau \ln\left(\frac{2q}{c\rho} \frac{e^{-\frac{R^2}{4a(\tau+\tau_o)}}}{4\pi a(\tau+\tau_o)\sqrt{4\pi a\tau}T} \right) \right]^{1/2} \quad (8)$$

Have some mathematical transformations and conditions the equation (8) can be written as follows:

$$\delta = N\sqrt{a \cdot \tau \cdot \ln\left[K \frac{D^3 \rho_1}{c\rho}(c_1 T + \lambda)\right]}$$ (9)

Where δ is the layer depth, m. D is the diameter of a drop of the sprayed aluminium or its alloy. τ is the time of heat diffusion into the polymer layer. A drop of sprayed aluminium with temperature T that contacts polymer layer generates the heat. This time τ is calculated with the equation (10):

$$\tau = (T \cdot c_1 \cdot \rho_1 \cdot V) / (\alpha \cdot S \cdot T_o)$$ (10)

Where V is the volume of a drop (m^3). S is the area of a drop surface, m^2. α is coefficient of heat rejection, W/ (m$^{2 \cdot o}$C). To is the temperature of a cooled drop ($T_o \approx 100°$C). N is a semi empirical coefficient that is in the range of 2.2-2.6. λ is the local heat to melt aluminium or its alloy, J/kg. ρ_1 is the density of aluminium or its alloy, kg/m^3. c_1 is the local thermal capacity of aluminium or its alloy, J/ (kg$\cdot^o$C). a is the thermal conductivity of the polymer layer. c is the heat capacity of the polymer layer, J/ (kg$\cdot^o$C). ρ_1 is the density of the polymer layer, kg/m^3. K is an empirical factor that equals to $1.15 \cdot 10^{11}$.

Table 1 lists data [8, 9] involved in equations (9). Obtained data shows an effect of diameter of sprayed aluminum drops and its temperature on minimal thickness of polymer layer that can be overheated by Al particles without catastrophic welding, degradation and failure of polymer. Results revealed that to prevent degradation of mechanical properties and to provide overall strength of the bearing, thickness of polymer layer should be at least 400 µm to form on it aluminium layer with the particles above 50 µm in diameter and about 1000 ^{0}C temperature.

Table I. Parameters of Al and polymer layers

Polymer type	Diameter of sprayed Al drops, µm	Temperature of sprayed Al drops, ^{0}C	Thickness of polymer layer, µm
Capron-polyamide	8 – 16	700 – 800	120
	15 – 40	750 – 900	175
	35 – 110	1050 – 1300	445
Polyethylene (dense type)	10 – 20	700 – 800	142
	15 – 35	750 – 900	271
	30 – 120	1050 – 1300	595

Analyzing the equation (9), it was revealed that the thickness of polymer layer is proportional to temperature and size of sprayed Al particles. For example, at thermal spraying of Al powder Al particles of 40-160 μm diameter at the 165-230 μm distance from substrate have the temperature ranged in 900-1700°C; at thermal spraying of Al cord the Al particles of 10-50 μm diameter at the 120-200 μm distance from substrate have the temperature ranged in 750-850°C.

MATERIALS, EQUIPMENT AND TEST TECHNIQUE

The base material of the plate sample was steel. Prior to thermal flame spraying of polymer surface of base metal was prepared by grinding with water jet contained SiC, Al_2O_3 particles with average size of 2 mm. In result, profile of the surface has been arranged in the quincunx order with roughness of 400 μm. Apply thermally sprayed bond coating to suit the application and to facilitate good chemical and mechanical bond between coating and core. Then grit blast and preheat surface. Air consumption was 0.4-0.45 m³/min. Distance of spraying is 120-140 mm. Polymer granular powder was used to coat surface with polyamide layer of 420 μm in thickness and 6 GPa in Young's modulus. The theoretical calculations by equations (1, 9) and practical approach show that aluminum layer formed by thermal flame spraying do not overheat polymer layer as temperature of Al particles do not exceed 120°C.

Some of technologies applied to produce aluminum such as Osprey technology /5/ use special expensive equipment. In contrast to Osprey, thermal flame spraying process is to be used as prototype to form aluminum layers with very close desired properties. In this way, preliminary task was to compare Young's modulus of thermally sprayed Al layer and the same layers produced by other processes. Young's modulus of sprayed layers had been calculated by technique described in [4] that is based on NDE ultrasound tests in tested samples.

To form aluminum layer was used both aluminum cord with diameter of 2 mm or aluminum powder with granules size of 60-100 μm and purity of 99.85%. Aluminum thickness was about 3 mm. Butane was a heating gas of the system to provide temperature about 1200°C for melting the aluminum powder. Thickness of the aluminum and polymer layers was calculated with equations [1, 2] respectively. By this way it provides optimal set of mechanical properties of aluminum layer. They are porosity 1-3%, Young's modulus 68-72 GPa, and fine structure.

Finally, aluminum layer was transformed into hard oxide aluminum with micro arc oxidizing process during processing in 60 min. Micro arc oxidizing have been done by special equipment described in /9/ at current frequency of 50 Hz, voltage of 420 V and current density of 18-20 A/dm². The electrolyte was based on distilled water with KOH (NaOH) of 3 gr/l and other additives to control pH, temperature, electrical conductivity et al.

The metallographic analysis of the cross-sectional micro-sections of the coated samples was made with a microscope. Microhardness was measured with Vickers indentation at load on the indenter of 0.5 N for 30 s. Then a diagonal of indentation track was measured and microhardness of the coatings was calculated. Porosity of the oxide hard coating was measured by the linear method (method of secant line). The system of texture image analysis "Leitz-TAS" (Germany) and optical microscope "Ortoplan" was used to study porosity of the oxide ceramic coatings.

RESULTS

Figure 1 shows microstructure of the coating. At the figure 1, hard alumina layer (line 1) is produced on soft composition of aluminum (line 2) and polymer layer (line 3) sprayed on steel base. The figure illustrates multilayered structure having arranged aluminum layer dispersed into the polymer layer. An analysis of structure and porosity in depth of alumina revealed that the structure of the layer contains α, γ phases of aluminum oxides, porosity ranges in 5-11 %. Microhardness of the layer reaches up to 15 GPa, that is very high result for alumina layer on sprayed aluminum in contrast to traditionally produced on cast aluminum alloys.

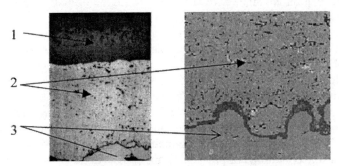

Figure 1. Structure of the composite

Considering data of Young's modulus [8, 9], thermally sprayed aluminum is found to be produced with an advanced Young's modulus in contrast to welded aluminum alloy with the same structure. There is two possible mechanisms to be considered in view of strength and modulus of elasticity of sprayed Al layers. On the one hand, energy of impact of flying Al particles on substrate destructs oxide film on aluminum particles due to its oxidation in a flight. On the other hand, in thermally sprayed aluminum layer processes of diffusion are intensified by high temperature of Al particles that reaches above $660\,^{\circ}C$. Observed intergranular diffusion and contact of particles improves modulus of elasticity of thermally sprayed aluminum in comparison with known prototypes based on welded alloys

[2, 3]. The sprayed layer contains up to 30 % of α, γ oxide phases of alumina that grounds the choice of sprayed aluminum and its technology to produce composite layers providing an advanced strength of the bearing.

RHEOLOGICAL MODEL DEVELOPMENT

To evaluate mechanical properties of composite systems there are many rheological models including integral elastic-tenacious-plastic systems [9, 13-16]. However, despite of some advantages known models do not completely describe mechanical behavior (law of deformation rate and stresses) of the composites like alumina-aluminum-polymer-steel substrate.

Based on recent researches of mechanical properties of alumina-based ceramics [14-16], polymers [13] and its composites the requirements to the rheological model have to be formulated to develop adequate model describing tight - strained state of the composite. Therefore, the following requirements have been found to be effectively used in rheological model describing composites' behavior:

1. since the composite include hard alumina layer and steel substrate that exhibit plasticity, the irreversible deformations have to be considered as plastic in nature. In plastic materials deformation develops only after loading by ultimate yield strength for the particular layer of the composite.

2. if deformation are smaller then yield strength, deformation at constant stress have to grow up step-by-step to a final stage;

3. under cyclic loading plastic deformation of the composite is summarized;

4. curve of deformation vs. time at constant load exhibits a linear dependence in one of a plotted region.

5. at unloading the retardation of deformations (elastic return) can be observed;

6. at constant deformations stresses are relaxed.

Along with above stated conditions, it is important to use the following mathematical requirements: (a) the system has to be solved to final decision (order of a required differential equation on stress and deformation should not exceed numbers of possible conditions of physical limitations); (b) the system and formulated problems have to be solved concerning stress or strain rates. Rational choice of the model that adequate to the studied composite materials will be defined by comparison of calculated and experimental results.

Based on above stated requirements describing aspects of mechanical behavior of the composite materials there can be found many rheological models, for example, for the elementary single dimensional deformation. However, to check an experimental adequacy of rheological models within suggested approach it is suggested to consider the following models.

The composite of hard alumina-aluminum can be presented as elastic-tenacious-plastic rheological model of the composite (fig. 2a). The mechanical prototype of the model is described in [17]. Structural equations of the integral model of the composite are (H || N || St-V) - (H-N || H). The kind of the rheological equation depends on a level and form of stress applied on model. The rheological model consisting from two elastic elements and one tenacious element (fig. 2b) can present the polymer layer. As a prototype of the model, it is considered the Maxwells' model and elastic element [13-16].

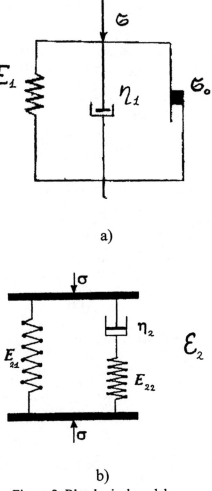

a)

b)

Figure 2. Rheological models

To simplify development and analysis of the system of differential equations it is taken advantage of the applicable rheological models. The selection of model that adequate to the studied composite material is determined by comparison of developed models and experimental results.

Integral deformation of sequentially connected models is calculate as follows:

$$\varepsilon = \varepsilon_1 + \varepsilon_2, \tag{11}$$

If the loaded system has the stress condition $\sigma > \sigma_0$, then the equation of the composite system (model 1 in fig. 2a) can be written by:

$$\sigma = \sigma_0 + E_1\varepsilon_1 + \eta_1 \frac{d\varepsilon_1}{dt} \tag{12}$$

where σ_0 σ, are stresses at initial and final time of the loading sequentially; E_1 is modulus of elasticity of the layer; η is coefficient of viscosity; ε is deformation.

For the second part of composite (model 2, fig. 2b) including polymer layers rheological differential equation is defined by:

$$\frac{d\sigma}{dt} + \frac{E_{22}}{\eta_2}\sigma = (E_{21} + E_{22})\frac{d\varepsilon_2}{dt} + \frac{E_{21}E_{22}}{\eta_2}\varepsilon_2 \tag{13}$$

where E_0, E_1 are modulus of elasticity of the material; η is coefficient of viscosity.

At random loading conditions the rheological equation of consequently connected system (model 1 + model 2) is given by:

$$\frac{\eta_1}{E_{22}}\frac{d^2\sigma}{dt^2} + \left(\frac{E_1}{E_{22}} + \frac{\eta_1}{\eta_2} + \beta\right)\frac{d\sigma}{dt} + \frac{E_1 + E_{21}}{\eta_2}\sigma - \frac{E_{21}}{\eta_2}\sigma_0 = \beta\eta_1\frac{d^2\varepsilon}{dt^2} + \left(\frac{\eta_1}{\eta_2}E_{21} + \beta E_1\right)\frac{d\varepsilon}{dt} + \cdot$$
$$\tag{14}$$

By mathematical transformations the equation (6.4) can be written as:

$$\sigma(t) = e^{-\frac{A}{2}\Delta t}\left[sh\left(\frac{AD}{2}\Delta t\right)\cdot\frac{x-y}{D} + xch\frac{AD}{2}\Delta t\right] + \sigma_0 - x$$
$$\tag{15},$$

where constants are:

$$x - y = \sigma_0 - \frac{S_0^{\cdot}}{B} + \frac{AS_1^{\cdot} - 2S_2^{\cdot}}{B^2} - \frac{2A^2S_2^{\cdot}}{B^3} - \frac{2\sigma_0}{A} + \frac{2S_1^{\cdot}}{AB}; \quad A = \frac{E_{22}}{\eta_1}\left(\frac{E_1}{E_{22}} + \frac{\eta_1}{\eta_2} + \beta\right);$$

$$B = \frac{(E_1 + E_{22})/E_{22}}{\eta_1\eta_2}; \quad D = \sqrt{1 - 4\frac{\eta_1}{\eta_2}\cdot\frac{(E_1 + E_{21})/E_{22}}{(E_1/E_{22} + \eta_1/\eta_2 + \beta)^2}};$$

$$S_{21} = \frac{E_1E_{21}E_{22}}{\eta_1\eta_2}\cdot\frac{1}{2}k_2$$

$$S_0 = \frac{E_{21}E_{22}}{\eta_1\eta_2}\sigma_0 + \beta E_{22}k_2 + \frac{E_{22}}{\eta_1}\left(\frac{\eta_1}{\eta_2}E_{21} + \beta \cdot E_1\right)k_1 \ ;$$

$$S_1 = \frac{E_{22}}{\eta_1\eta_2}\left(\frac{\eta_1}{\eta_2}E_{21} + \beta \cdot E_1\right)k_2 + \frac{E_1 E_{21}E_{22}}{\eta_1\eta_2}\varepsilon_0 \ .$$

Using mathematical transformations involving parameters of the Hertz's' theory, the system of the equations is found to be solved in respect of applied load and acting stresses in the composite. Function of stress-deformation mode of the composite system in fig. 2a is defined by equation (12) at linear function of deformation rate vs. loading time. Function of stress-deformation mode of the composite system in fig. 2b will be expressed by integrating equation (14) with some conditions for Hertzian contact of a sphere and a plate at linear function of deformation rate vs. loading time. Constant parameter will then read as $C=0$ in equation (14), that is taken to condition of $\sigma(0) = 0$. In result function of stress-deformation mode is defined by:

$$\sigma(t) = E_{21}[\varepsilon(t) - \varepsilon_0] + \eta_2\dot{\varepsilon}(t) - \frac{3k\eta_2}{2t_1}e^{-t/\tau} + \frac{k\eta_2^2}{E_{22}t_1^2}\left(1 - e^{-t/\tau}\right) \qquad (16)$$

where $\tau=\eta_2/E_{22}$. $t_1=R/V$; $k=4/(3\pi\cdot(1-\mu^2))$.

Difference in the squared brackets ε and ε_0 is derived from equation (14) and is described deformations resulted by loading. Here of interest is summarized deformation that equals to sum of initial deformation of rough surface layer and deformation resulted by loading.

The above equations define the functions of stress-deformation mode expressing rheological behavior of complex composite system, however its hand-made application could be very hard procedure due to complex mathematical and analytical calculations to be done. Moreover, under some conditions of loading simple rheological models (fig. 2a,b) can show adequate results that will then well fitted to experimental base. So, in this work equations (12, 16) will then be used in experiments. To reveal a function of contact stress rate, stress function $\sigma(t)$ is multiplied with actual area of the indentation track area as it follows $S=\pi d^2(t)/4$.

In result the function can be written as

$$P(t) = \sigma(t)\cdot\pi d^2(t)/4, \qquad (17)$$

or by accepting conditions of Hertzian contact and the rheological equations, the function of contact stress rate is modified to

$$P(t) = \sigma(t)\left[\pi RVt + \frac{\pi d^2(t)}{4}\right], \tag{18}$$

where R is indenter radius, V is velocity of indenter loading, t is a time of loading, $d(t)$ is function of indentation track depth vs. applied load ($d(t) = 2\sqrt{R\alpha(t)}$), $\sigma(t)$ is the function of stress rating at applied load, $\varepsilon(t)$ is the function of deformation rate ($\varepsilon(t) = 2\,kd(t)/R$).

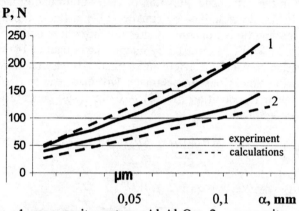

1 – composite system «Al-Al$_2$O$_3$», 2 – composite system "steel – viscous-elastic material – aluminum - alumina".

Figure 3. Indentation depth vs. load

Figure 3 shows indentation depth vs. load curve. It was plotted as mechanical response of the composite under Hertzian indentation at constant rate with sintered alumina ball of 4 mm diameter. The composite exhibits linear relation of stress curve. At unloading (not shown in fig.3) the composite could show retardation of deformations (elastic return) because of damping viscous-elastic material. It could be shown in a figure as a downfall curve. The plotted relations of experimental data and calculated data (equations 12, 16, 18) are showed a good agreement of developed rheological model and mechanical behavior of the composite. The above stated conditions are found to be used in investigations of mechanical and rheological properties of "alumina-aluminum- viscous-elastic material (polymer) – steel" composite systems.

CONCLUSION

Applied soft intermediate composite layer of aluminum and polymer provide both rigid and adaptive structure of the composition that, however, is deformed under applied load, but preventing the hard coatings from ultimate failure. In result, the composition has both strength and the ability to adapt under applied localized contact pressure.

Thermal spraying process is found to be effective to produce polymer-aluminum composition to be used to form alumina layer by micro arc oxidizing. Equations to calculate effective thickness of thermally sprayed aluminum and polymer layers have been developed. It was found that of interest are temperature and size of sprayed Al particles, temperature of polymer surfaces and consumption of spraying air. Varying the parameters elasticity modulus of sprayed aluminum can reach up to 75 MPa.

Experiments show that thickness of polymer layer should be at least 400 µm to form on it aluminum layer with Al particles above 50 µm in diameter and its 10000C temperature. Revealed technological regimes and equations to calculate thickness of polymer and Al layers give an approach to an engineering design of the composite in view of Young's modulus, adhesion and strength. Structure of alumina consists of α, γ phases of aluminum oxides. Porosity of alumina ranges in 5-11 %. Microhardness of alumina reaches 17 GPa.

Several requirements to rheological modeling of composite systems have been revealed. Rheological models to be applied to the composite systems have been proposed for major cases of applied loading mode and have been confirmed by in-suit experiments using Hertzian theory for spherical indentation. Applying the rheological models in analysis of rheological behavior of the composite systems it is possible to find correlation between rheological properties (plasticity, elasticity and tenacious), microstructure and composition of the systems to analyze its strength and load rating characteristics. Rheological models with Hertzian theory gives reliable explanation of the contact at loads up to 500 N for investigated composite systems.

Virtually in ideal case, of interest to be considered is adaptive ability of the construction in which the polymer-aluminum composition or what ever plays key role. Since the paper highlights only top of our work we expect to present our project in progress idea in future works.

ACKNOWLEDGEMENTS
This work was performed under the scientific direction and principal supervision of Dr. Sc., Professor Zinovii P. Shulman (Institute of Heat and Mass-transfer, National Academy of Sciences of Belarus) and Dr. Sc., Prof. P.A.Vitiaz (INDMASH NAS of Belarus). Special thanks for help in rheological modeling of the composites to A.A.Mahanek, PhD (ITMO NAS of Belarus).

REFERENCES

[1]Jerry L. Patel, Nannaji Saka, Ph.D. A new Coating Process for Aluminum. Microplasmic Corp., Peabody, Mass. www.microplasmic.com, (2000)

[2]DE № 3934141, F 16 C 33/12, 1990.

[3]DE № 4038139, F 16 C 33/10, 1990

[4]Wiktoreks S. Studies of physical properties of hot sprayed aluminum metal coatings to steel substrates. Fnti. Corrosion. 33, 1986, p. 4-9.

[5]Lavernia E.J., Grant N.J. Spray deposition of metals: a review // Material Science and Engineering. – 1988, 98, p. 381- 384.

[6]Willis T. Spray deposition process for metal matrix composites // Metals and Materials. – 1988, 4, p. 485 – 488.

[7]Friction and wear behaviors of alumina-based ceramics in dry and lubricated sliding against tool steel. Yang C. "Wear", 1992, 157, №2, p.263-277

[8]Berestnev O.V., Basenuk V.L., Kireytsev M.V. et al. Composite coating and technology of its manufacturing. C 23 C 28/00. Patent RU №2000111046/02, 2001.

[9]Basenuk V. ed. in chief, Kireitsu M. et al. "Development of fundamentals for technology and design of composite powder materials based on aluminum with non-metallic components and formation of oxide ceramic coatings on the base of the materials by the electrochemical method of micro arc oxidizing for an application in the hard loaded friction pairs and bearings". Report under the theme "Material-65". Registration Number № 19962944. INDMASH NAS of Belarus, Minsk, (2000), p. 165.

[10]Kireytsev M.V. Mechanical properties of composite coatings based on hard anodic oxide ceramics. // Proc. of the SEM Annual Conference and Exposition. Best Student Paper Competition. June 4-6, Portland, Oregon, USA (2001) pp.12-13

[11]Kireitseu M.V. and Basenuk V.L. Study of Load Rating of CrC–Al2O3 –Al Composite Coating. // Proc. of 2001 TMS Fall Meeting: "2nd Int. symposium on modeling the performance of engineering structural materials (MPESM – II)", Ed. Dr. Lesurer. – 18-21 October, Akron, OH, USA, 2001. – P. 355 – 364.

[12]Kireitseu M.V. and Basenuk V.L. Investigation of tribomechanical properties of Al–Al2O3 and Al–Al2O3–CrC composite coatings based on the oxide ceramics. // Proc. of 2001 TMS Fall Meeting: "2nd Int. symposium on modeling the performance of engineering structural materials (MPESM – II)", Ed. Dr. Lesurer.

– 18-21 October, Akron, OH, USA, 2001. – P. 365 – 377.

[13]Rudnitski V.A., Kren A.P. and Shilko S.V. Rating the behavior of elestomers by their indentation at constant rates. Wear, Vol. 22, №5, Sept-Oct. 2001, pp. 502-508.

[14]Basenuk V., Kireytsev M., Fedaravichus A.. Application of rheological models to composite sliding bearings based on polymer-metal-ceramics composition. // Proc. of the Symposium of Material&Construction Failure.23-25 May01 Augustov,Poland. pp.12.

[15]Kireitseu M.V. Fatigue of composite coatings based on hard oxide ceramics and chrome carbide. // Proc. of the SEM Annual Conference. Session 39, Paper #87. – June 10–12, Milwaukee, Wisconsin, USA, 2002. – P. 265–271

[16]Kireitseu M.V. and Yerakhavets L.V. Fracture of anodic hard ceramic–based composite coatings on a metal substrate. // Proc. of the AESF SUR/FIN conference. Ed. Dr. Dick Baker. – 24-27 June, Chicago, IL, USA, 2002. – P. 96–104.

[17]Shulman et al. Rheological behaviour of the composite systems and structures. Published by "Nauka", Minsk, 1978, p.378.

Processing–Microstructure–Property Relationships

MONOCLINIC-TO-TETRAGONAL TRANSFORMATION AND MECHANICAL PROPERTY RECOVERY IN LOW TEMPERATURE WATER-DEGRADED 3Y-TZP PROCESSED BY A CARBURIZING TREATMENT

Zhenbo Zhao   Cheng Liu*   Derek O. Northwood [+]

Mechanical, Automotive & Materials Engineering
University of Windsor, Windsor, Ontario, Canada   N9B 3P4

* Mechanical, Aerospace & Industrial Engineering
Ryerson University, Toronto, Ontario, Canada, M5B 2K3

[+] Also Faculty of Engineering & Applied Science
Ryerson University, Toronto, Ontario, Canada M5B 2K3

ABSTRACT
      The reverse transformation (from monoclinic to tetragonal phase) by carburizing at different temperatures and mechanical property recovery in a low temperature water-degraded 3Y-TZP ceramic was investigated. The effect of the introduction of carbon on the monoclinic-to-tetragonal transformation for the aged 3Y-TZP ceramic was compared with that for a pure annealing process (1200°C for 1 hr). The stronger stability found by carburizing than by pure annealing is attributed to the combined effects of anion- and cation-stabilizers, carbon and yttrium ions, since carbon ions occupy mainly the octahedral interstitial sites, rather than yttrium ions, which stabilize TZP by creating oxygen vacancies to maintain charge neutrality after substitution of zirconium in TZP. It is believed that their effects on the stabilization of zirconia are additive due to the different stabilizing mechanisms and positions (carbon for surface, yttrium for bulk). For aged 3Y-TZP, the degraded mechanical properties bending strength and fracture toughness, can be effectively recovered by carburizing.

INTRODUCTION

Since the first report by Masaki et al[1], the low temperature environmental degradation phenomenon of Y-TZP (yttria-stabilized tetragonal zirconia polycrystals) has been extensively investigated. The various mechanisms of low temperature environmental effects on transformation-toughened zirconia ceramics have been recently reviewed by Zhao and Northwood[2]. Although many methods[3-10] have been successfully implemented to prevent the low temperature degradation behavior, there have been no efforts to recover Y-TZP's mechanical properties after the low-temperature degradation. Due to the unique stabilizing mechanism for carbon in TZP as reported by Zhao, Liu and Northwood[7, 10], the reverse transformation (from monoclinic to tetragonal phase) by carburizing at different temperatures in a low temperature water-degraded 3Y-TZP ceramic was investigated. If successful, this method will make mechanical property recovery and the related crack healing of a low temperature water-degraded 3Y-TZP ceramic possible.

EXPERIMENTAL METHODS

Commercial raw zirconia powders containing 3mol% Y_2O_3 were used. They were uniaxially pressed at 150MPa in a rectangular mold to form rectangular samples 4.5 by 5.5 by 38mm and 4.5 by 5.5 by 42 mm. The samples were sintered at 1600°C for 3 hrs in air, and then cut, ground and polished to a final specimen size of 3 by 4 by 36 mm for the bending strength tests, and 3 by 4 by 40 mm for the fracture toughness tests. The bending strength was obtained by 3-point bending tests with a span of 30 mm and a cross-head speed of 0.5 mm/min. The fracture toughness was determined by the chevron-notched beam technique. The chevron notches were machined with a 0.25-mm-thick diamond saw. The notched specimens were annealed at 1000°C in air for 1.5 hours to relieve the residual stress introduced by the saw. Fracture toughness tests were performed by four-point bending using a cross-head speed of 0.05 mm/min, and K_{IC} values were calculated using the equation reported by Munz *et al* [11]. A high sintering temperature of 1600°C was intentionally used to obtain t-ZrO_2 with a large grain size, which is prone to phase transformation and, thus low-temperature degradation. Some selected samples were annealed in a hydrothermal corrosion environment (in an autoclave with water) at 200°C at 1.2 MPa up for to 400 hours in order to produce the low-temperature degradation of 3Y-TZP. The low temperature water-degraded 3Y-TZP samples with a mirror surface were further heat treated by a carburizing process in which the samples were buried in graphite powder at 1200°C to 1600°C for 2 to 8 hours as describedin ref. [7]. Also, the low temperature water-degraded 3Y-TZP samples were annealed at 1200°C in air only for 2hrs in order to compare them with the carburizing treatment The test for the low temperature degradation was performed at 200°C in water at 1.2MPa up for to 400h for both the surface carburized and pure annealed 3Y-TZP specimens in order to compare their stability.

Phase analysis was carried out by x-ray diffraction techniques for all samples. The fraction of monoclinic phase on the surface was determined by Garvie's method [12] as follows:

$$X_m = [I_m(111) + I_m(11\bar{1})] \ / \ [I_m(111) + I_m(11\bar{1}) + I_c(111)] \tag{1}$$

where, $I_m(111)$ and $I_m(11\bar{1})$ are the intensity of the monoclinic (111) and $(11\bar{1})$ line, $I_c(111)$ is the intensity of cubic (111) line in the X-ray diffraction (XRD) pattern. The bulk density of each sample was measured using Archimedes method. The average grain size was determined by scanning electron microscopy (SEM) and the line intercept method [13]. For the SEM observations, the grain boundaries were etched in air at 1450°C for 0.5 hour.

RESULTS AND DISCUSSION

The XRD patterns show that there is almost 70% monoclinic-phase existing in the surface of 3Y-TZP samples after a 400 hrs water degradation treatment at 200°C (see Fig. 1 A). After the carburizing treatment at 1600C for 4 hrs for the water-degraded 3Y-TZP, 80% of the m-phase can be transformed back to the tetragonal phase as shown in Fig. 1 B. This agrees well with our previous results[7, 10] regarding the unique effects of carbon on the stability of 3Y-TZP.

SEM shows that as-sintered 3Y-TZP has a relatively narrow grain size distribution as shown in Fig. 2. An average value of 0.8μm was obtained by the line intercept method [13].

The average grain sizes for the water-degraded 3Y-TZP after carburizing were compared with that of pure annealing process treatment. Table 1 shows the effect of temperature on grain growth after the different processes. An increase in grain size with increasing carburizing temperature is not surprising, and is typically observed. The 3Y-TZP with a large grain size is prone to phase transformation and, thus to low-temperature degradation[2]. This is shown by the higher monoclinic phase content of the water-degraded 3Y-TZP after a pure annealing treatment at 1200°C (0.97μm grain size) than that of 3Y-TZP (0.8μm grain size) as shown in Fig. 3. This also explains why the bending strength of water-degraded 3Y-TZP after the 1600°C carburizing is lower than that of the water-degraded 3Y-TZP after carburizing at 1500°C due to the fast grain growth at 1600°C (see Fig. 4), and also why the fracture toughness increases after a pure annealing treatment as shown in Fig. 5 due to the slight increase in grain size.

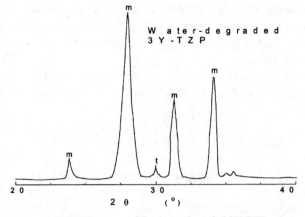

A. XRD pattern of the water degraded 3Y-TZP

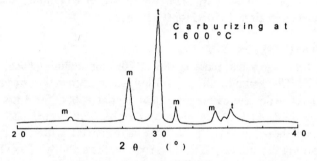

B. XRD pattern of the water-degraded 3Y-TZP after carburizing at 1600C for 4 hrs

Fig. 1 XRD patterns of 3Y-TZP samples after different treatments

Fig. 2 SEM micrograph of 3Y-TZP after sintering at 1600°C for 3 hrs

Table 1 Comparison of the average grain size for water-degraded 3Y-TZP after carburizing and pure annealing treatments.

Method of treatment	Carburizing					Pure Annealing
Temperature of treatment	1200°C	1300°C	1400°C	1500°C	1600°C	1200°C
Grain size (μm)	0.96±0.14	1.01±0.20	1.25±0.12	1.72±0.19	3.21±0.22	0.97±0.16

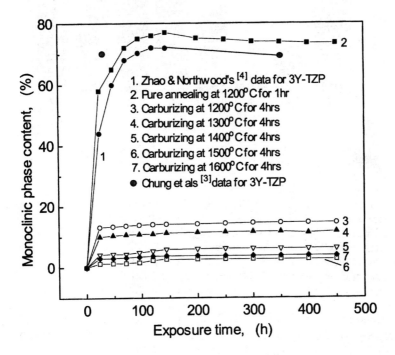

Fig. 3 Comparison of the m-phase content for 3Y-TZP and water-degraded 3Y-TZP after different heat treatments before and after a hydrothermal treatment at 200°C and 1.2MPa for different exposure times.

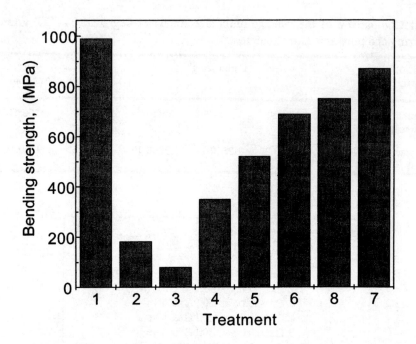

1. Zhao & Northwood's[4] data for 3Y-TZP
2. Zhao & Northwood's[4] data for 3Y-TZP after the low temperature hydrothermal treatment
3. Pure annealing at 1200°C for 1 hr for the water-degraded 3Y-TZP
4. Carburizing at 1200°C for 4 hrs for the water-degraded 3Y-TZP
5. Carburizing at 1300°C for 4 hrs for the water-degraded 3Y-TZP
6. Carburizing at 1400°C for 4 hrs for the water-degraded 3Y-TZP
7. Carburizing at 1500°C for 4 hrs for the water-degraded 3Y-TZP
8. Carburizing at 1600°C for 4 hrs for the water-degraded 3Y-TZP

Fig. 4 Bending strength of 3Y-TZP and water-degraded 3Y-TZP after pure annealing and carburizing treatments before and after a hydrothermal treatment at 200°C and 1.2MPa for 400 hrs.

The effect of the exposure time in the second hydrothermal treatment on the monoclinic phase content for materials after either carburizing at different temperatures or a pure annealing treatment are compared in Fig. 3. In contrast to 3Y-TZP[3, 4], which shows poor resistance to hydrothermal corrosion (more m-phase is formed), a good resistance of the water-degraded 3Y-TZP to hydrothermal corrosion was found after carburizing at all temperatures. The x-ray diffraction results show that there is almost no further m-phase formed in the surface of the carburized 3Y-TZP(water-degraded) samples after hydrothermal treatments up to 400 hrs. The m-phase that was identified at the surface of same carburized 3Y-TZP samples was due to the incomplete reverse transformation upon carburizing, rather than by the hydrothermal treatment.

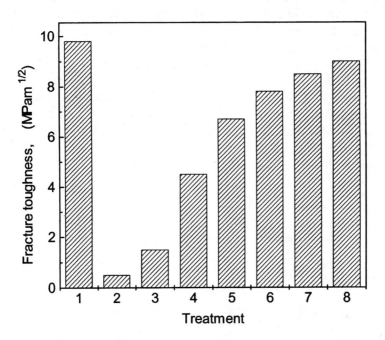

1. Zhao & Northwood's[4] data for 3Y-TZP
2. Zhao & Northwood's[4] data for 3Y-TZP after the low temperature hydrothermal treatment
3. Pure annealing at 1200°C for 1 hr for the water-degraded 3Y-TZP
4. Carburizing at 1200°C for 4 hrs for the water-degraded 3Y-TZP
5. Carburizing at 1300°C for 4 hrs for the water-degraded 3Y-TZP
6. Carburizing at 1400°C for 4 hrs for the water-degraded 3Y-TZP
7. Carburizing at 1500°C for 4 hrs for the water-degraded 3Y-TZP
8. Carburizing at 1600°C for 4 hrs for the water-degraded 3Y-TZP

Fig. 5 Fracture toughness of 3Y-TZP and water-degraded 3Y-TZP after pure annealing and carburizing treatments before and after a hydrothermal treatment at 200°C and 1.2MPa for 400 hrs.

The bending strengths and fracture toughness of 3Y-TZP and the water-degraded 3Y-TZP after pure annealing and carburizing treatments before and after hydrothermal corrosion are compared with 3Y-TZP in Fig.s 4 and 5. It was found that the degraded mechanical properties, bending strength and fracture toughness for aged 3Y-TZP, can be partially recovered by carburizing depending on the treatment temperature. Although the initial strength of 3Y-TZP is high enough for commercial application, the low strength of 3Y-TZP after annealing precludes its potential application in hydrothermal environments. A pure annealing treatment at 1200°C could not prevent the further low temperature degradation (see curve #2 in Fig. 3) even though the reverse transformation from monoclinic to tetragonal phase of the water-degraded 3Y-TZP can occur at temperatures above 750°C[14]. According to ZrO$_2$-Y$_2$O$_3$ phase diagram summarized by Yoshimura[15](see Fig. 6), the discrepancies in the precise location of the equilibrium temperature T$_0$ are due to the "non-ideal" distribution of the oxygen vacancies as well as to the non-ideal dopant concentration. However, the retransformation temperature from the monoclinic to the tetragonal phase for 3Y-TZP in the equilibrium state definitely occurs below 1000°C.

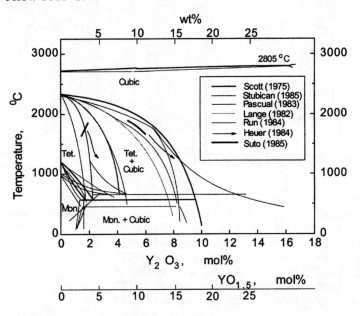

Fig. 6 ZrO$_2$-Y$_2$O$_3$ phase diagram (ZrO$_2$-rich region) as summarized by Yoshimura[15].

Advances in Ceramic Matrix Composites VIII

The reasons that carburizing not only can make the water-degraded 3Y-TZP reversed transformed from m- to t-phase, but also keep its t-phase stability for the further low temperature environmental treatment should be attributed to the unique stabilized mechanism of carbon.

Zhao, Liu & Northwood[7] have found that carbon mainly exists in the octahedral interstitial sites in the TZP lattice as pure atoms, rather than forming a Zr-O-C structure similar to the Zr-O-N or Zr-Y-O-N structures after nitrogen is introduced into the TZP lattice. No new vacancies are created since the access of carbon atoms to ZrO_2 lattice did not bring extra electrons or vacancies. However, Zhao Liu and Northwood[7] did not provide a detailed explanation of why carbon can stabilize 3Y-TZP after it occupies the octahedral interstitial sites in the TZP lattice. It well accepted that the internal stresses caused by the addition of cation-stabilizers (such as yttria in 3Y-TZP) is the necessary condition to retain the tetragonal phase at room temperature. The occupation of interstitial sites by any inclusion atom will cause lattice distortion. Therefore, the lattice distortion of 3Y-TZP resulting from the diffusion of carbon into the octahedral interstitial sites in the TZP lattice will result in a strain-stress field to compensate for the possible loss of internal strain and stress due to adsorption of water at the surface. This agrees well with Schmauder's[16] analysis that the thermal expansion difference between the a- and c- axes of the t-phase in $2mol\%Y_2O_3$ was greater than that of $3mol\%Y_2O_3$ resulting in a change in internal shear stress field, which could increase the driving force and thus cause the transformation.

CONCLUSIONS

1. The reverse transformation, from the monoclinic to the tetragonal phase, can be obtained by both carburizing at various temperatures and pure annealing at 1200°C.

2. The loss of mechanical properties in a low temperature water-degraded 3Y-TZP ceramic can be recovered by carburizing, rather than by pure annealing.

3. The stronger stability found by carburizing than by pure annealing is attributed to the unique stabilizing mechanism of carbon since carbon ions occupy mainly the octahedral interstitial sites, rather than yttrium ions, which stabilize TZP by creating oxygen vacancies to maintain charge neutrality after substitution for zirconium in TZP.

4. This method facilitates the mechanical property recovery and related crack healing.

REFERENCES

[1]O. T. Masaki, H. Kuwashima, and K. Kobayashi, "Phase Transformation and Change of Mechanical Strength of ZrO_2-Y_2O_3 by Aging," in Abstracts of the

Annual Meeting of the Japanese Ceramic Society, A-3, Japanese Ceramic Society, Tokyo, 1981.

[2]Z. Zhao and D.O. Northwood, "Mechanisms of Low Temperature Environmental Effects on Transformation-Toughened Zirconia Cermics," Ceramic Engineering and Science Proceedings, American Ceramic Soc., 20[4] 95-103 1999.

[3]T. Chung, H. Song, G. Kim, and D. Kim, "Microstructure and Phase Stability of Yttria-doped Tetragonal Zirconia Polycrystals Heat Treated in Nitrogen Atmosphere," J. Am. Ceram. Soc., 80 [10] 2607-12 (1997).

[4]Z. Zhao and D.O. Northwood, "Combined Effects of Mullite and Alumina on the Stability and Mechanical Porperties of a Y-TZP Ceramic in a Hydrothermal Corrosion Environment," Ceramic Transactions, Vol. 103 , Advances in Ceramics-Matrix Composites V, Am. Ceram. Soc., 1999, pp. 503-514.

[5]Z. Zhao, and D. O. Northwood, "The Effect of Surface GeO_2-Doping and CeO_2-Doping on the Degradation of 2Y-TZP Ceramic on Annealing in Water at 200°C," Materials & Design, 20, [6] 297-301 (1999).

[6]Z. Zhao, C. Liu and D. O. Northwood, "Surface Modification of Yttria-Stabilized Tetragonal Zirconia Polycrystal/Alumina Composites by Incorporation of Mullite as a Second Phase," Ceramic Engineering and Science Proceedings, American Ceramic Soc. 21[3] 619-626 (2000).

[7]Z. Zhao, C. Liu and D. O. Northwood, "Carburizing of Tetragonal Zirconia," J. Aust. Ceram. Soc., 36[1] 135-139 (2000).

[8]T. Sato, S. Ohtaki, T. Endo, and M. Shimada, "Improvement to the Thermal Stability of Yttria-Doped Tetragonal Zirconia Polycrystals by Alloying with Various Oxides," in Advances in Ceramics, Vol. 24A, Science and Technology of Zirconia III, Edited by S. Somiya, N. Yamamoto, and H. Hanagita, American Ceramic Society, Westerville, OH, 1988, pp.29-37.

[9]Y. Koh, Y. Kong, S. Kim, and H. Kim, "Improved Low-Temperature Environmental Degradation of Yttria-Stabilized Tetragonal Zirconia Polycrystals By Surface Encapsulation," J. Am. Ceram. Soc., 82 [6] 1456-58 (1999).

[10]Z. Zhao, C. Liu and D. O. Northwood, "Carbonitriding of Tetragonal Zirconia," Ceramic Engineering and Science Proceedings, American Ceramic Soc. 22[4] 59-66 (2001).

[11]D. Munz, R. T. Bubsey, and J. L. Shannon, Jr., "Fracture Toughness Determination of Al_2O_3 Using Four-Point-Bend Specimens with Straight-Through and Chevron Norches," J. Am. Ceram. Soc., 63 [5-6] 300-305 (1980).

[12] R. C. Garvie and P. S. Nicholson, "Phase Analysis in Zirconia Systems," J. Am. Ceram. Soc., 55[6] 303-305 (1972).

[13]R. L. Fullman, "Measurement of Particle Size in Opaque Bodies," J. Metal. Trans., AIME, 197[3] 447-52 (1953).

[14]J. K. Lee, H. Kim, "Monoclinic-to Tetragonal Transformation and Crack Healing by Annealing in Aged 2Y-TZP Ceramics," J. Mater. Sci. Lett., **12**, 1765-1767 (1993).

[15]M. Yoshimura, "Phase Stability of Zirconia," Ceramic Bulletin, **67**[12]1950-1955 (1988).

[16]S. Schmauder and H. Schubert, "Significance of Internal Stresses for the Martensitic Transformation in Yttria-Stabilized Tetragonal Zirconia Polycrystals During Degradation," J. Am. Ceram. Soc., **69**[7]534-540 (1986).

HARD OXIDE CERAMIC COMPOSITE MODIFIED BY ULTRA-DISPERSED DIAMONDS

Maksim Kireitseu
Institute of Machine Reliability
National Academy of Sciences of Belarus
Lesnoe 19 – 62, Minsk 223052, Belarus

ABSTRACT

Hard alumina coatings containing ultra dispersed diamonds nanoparticles have been investigated by in suit experiments revealing structural properties of the composite. Results discovered effects of diamonds on roughness, microstructure and hardness of the alumina-based composites. Some technological parameters involved in the micro arc oxidizing processing of thermally sprayed aluminum-diamond substrate are discussed. Tribological behavior of the alumina-diamonds composite has been reviewed.

INTRODUCTION

The use of hard ceramic coatings on aluminum and/or polymeric composite components has been around for several years, particularly in the automobile industry where many exterior trim parts are produced by electroplating onto an etched polymer composite or substrate [1]. Similar components are widely being fabricated for machine building industry, power plant and decorative applications as well. Aluminum improves weight of structures and corrosion resistance at low costs of application. Hard alumina layer improves wear resistance of aluminum. An oxide film on Al particles provides advanced strength. However, such new types of coating-substrate system as hard coating on soft substrate has the need for improved load rating, wear rate and durability of the composition in general.

PROBLEM

In practice, engineers indicated [1, 2] that localized load and stress applied to relatively thin alumina layer induce its peeling off and significantly limit area of their application. Engineers concentrate attention on better strength and hardness of the coatings to be used in a machine.

One approach used to improve structure and mechanical properties of alumina layer is to control technological regimes and environment of electrolyte. Another one is to strength alumina by smart particles that gives composite material [2, 3]. Studies of showed that reinforcement by hard particles improves microstructures and mechanical properties.

It is well known that ultra-dispersed diamonds (UDD) are a new type of synthetic diamond powders. In some cases, UDD are produced by chemical purification of explosion products. As result, particles of UDD have spherical and isometric form without crystalline facets and a fractional structure of clusters. It is a high-dispersed powder with active surface. It is expected that ultra dispersed diamonds will improve mechanical properties of alumina coating formed on aluminium or its alloy. The paper outlines some results of the work.

OBJECTIVE OF RESEARCH

The objective of the work is to reveal an effect of ultra dispersed diamonds on microstructure, mechanical properties and friction force of alumina coatings.

MATERIALS AND EQUIPMENT

The base material of the plate sample was steel (45 HRC). Prior to thermal spraying of Al-diamonds-containing composite layer, base surface had been grinded by water jet contained SiC, Al_2O_3 particles of 1-2 mm size. Then grit blast and preheat surface. To produce aluminum layer, composition of diamond nanoparticles (fig. 1) and aluminum powder with Al granules size of 60-100 μm and purity of 99.85% was used. Concentration of diamonds in the powder was 0.1-0.15% of Al volume. Air consumption was about 0.5 m^3/min. Distance of spraying is about 200 mm. Thickness of the composite layer was about 4 mm. Butane was a heating gas of the system to provide temperature about 1200^0C for melting the aluminum powder.

The ultra dispersed diamonds (fig.1) of $60 \cdot 10^{-4}$ μm (60 Å) in average size have been added in aluminum powder. Diamonds were synthesized in strong non-equilibrium conditions of a detonation surge. The diamonds look like isometric fragments of three different shapes that are shown in fig. 1. Additional finishing of the particles can produce the particular shapes of diamonds. Table 1 lists major characteristics of ultra dispersed diamonds.

Composite aluminum-diamonds layer has been transformed into hard oxide aluminum with micro arc oxidizing process during processing for 60 min. Micro arc oxidizing technology uses current frequency of up to 1000 Hz, voltage up to 450 V and alternating current density of 13-20 A/dm^2. The electrolyte was based on distilled water with 3 gr/l KOH or NaOH, and other additives to control pH, temperature, electrical conductivity of the electrolyte etc. Some papers [1, 2, 8] refer to detailed description of the process. Finally, the alumina-based coatings of 150 μm thickness have been produced.

The alumina layer was produced by special equipment and electrolyte that transform sprayed Al layer with UDD particles into hard alumina. The equipment provides either processes electroplating or micro arc oxidizing. Figure 2 shows the principal structural scheme of the equipment. The system has the intelligent controller that consists of control unit (CU), control desk (CD), filter unit (FU) and one or two-power supply unit (PSU). The used power supply worked in the anodic-cathode regime on high frequencies of current that has special form of periodic impulses [8, 9]. All devises are connected to the main computer (CC). Read-only memory (ROM) stores program of the process. Random-access memory (RAM) records discrete information through the interface module (IM). In result the process regimes can be changed cyclically in accordance with defined program, using direct current (DC), reverse current (RC), pulse current (PC), programmable regime (PrR) with positive and negative impulses of current at various frequencies bands.

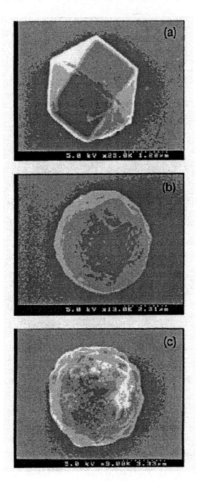

Figure 1. Diamond nanoparticle of different shapes

Table I. Characteristics of diamond particles

Characteristics of ultra dispersed diamonds particles	
Major chemical composition of UDD particles	82 – 92 % carbon, 1 – 3 % of nitrogen, 1 – 2% of hydrogen, and up to 1 % of other additives.
Phase composition of UDD	80 – 100 % of cubic diamond, 0 – 10 % of hexagonal diamonds, and up to 20 % of diamond as X-ray amorphous carbon.
UDD particles size	4 to 8 nm
Aggregated clusters	20 - 30 nm
Surface area	300±30 m³/gr
Density	3.1 - 3.2 gr/cm³
Thermo-stability until oxidation	400 - 450 °C
Thermo-stability until graphitization	1000 - 1100 °C

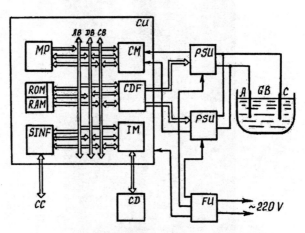

Figure 2. Power supply and equipment for micro arc oxidizing/electroplating

TEST TECHNIQUE

Morphology of the samples was investigated by the scanning electron microscope (SEM). Crystalline structure of the alumina films was analyzed by X-ray diffractometer. Microstresses, grain size and orientation index (texture degree) were calculated by using well-known methods highlighted in [6, 7]. Electron Spectroscopy for Chemical Analysis (ESCA) identified composition of thin films. Roughness and microrelief of the coatings were measured by the profilograph-profilometer.

Microhardness was measured with Vickers indentation at load on the indenter of 0.5 N for 30 seconds. Then a diagonal of indentation track to be measured and microhardness of the coatings were calculated.

In this work, the Pin-on-Disk Tribometer as a perfectly suited tool for characterizing the tribological characteristics of alumina-based composite has been used an expected feature in various applications. Of interest is the fact that the coating consists of a very hard alumina matrix with deposited hard diamonds into the matrix. In general, scratch testing would not be recommended for this kind of sample, as the small hard diamond tips usually used for this test would fail much earlier than the coating, causing inaccurate results, that has to be adjusted.

The testing procedure consists of a loaded static partner (standard pin of 6 mm^2 contact square) in sliding contact with the coated disk sample. The entire contact zone has been investigated in a dry contact, in a distilled water lubricant or machine oil. Applied load was constantly increased at 10 N for 10 m of sliding distance, if no significant changes in frictional behavior have occurred. The friction force can be accurately monitored until a sharp decrease signal that the coating has failed and that the substrate has been reached. On the other hand, the friction process stabilizes that gives actual friction force and coefficient of friction as well. The actual failure mechanism can be confirmed by optical microscope observation of the wear track at various stages during the test.

RESULTS & DISCUSSION

Result of investigations revealed the following.

1. Ultra dispersed diamonds effects on roughness, microstructure and hardness of the alumina-based composites.

Microstructure of the composite became fine with presence of UDD particles. Roughness decreases up to Ra 0.5-0.3 μ m Average grain diameter decreases from 15-25 nm for single alumina layer to 1-2 nm for alumina formed for Al-diamonds substrate. UDD crystals are distributed into alumina-based matrix. In a result, many crystallized centers and clusters of particles are observed at the surface of the coating in Fig. 3 and 4. The alumina structure without UDD particles has axial texture. Grains orientate mainly in <111> direction. Hardness of the composite with UDD particles increases up to 26-27 GPa at 0.1% UDD concentration.

Researches showed that ultra dispersed diamond particles were involved into the polycrystalline alumina-based matrix. The presence of diamond particles in the composite structure was indicated by EDS analysis near the surface of the film, as shown in the figures 5 and 6. On surface of the coating can be observed strong pudding rocks of fragments based on diamonds. Clusters and separate UDD particles conglomerates with alumina hard phases that are produced during substrate oxidizing.

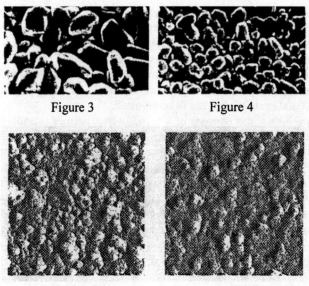

Figure 3   Figure 4

Figure 5   Figure 6

2. Diamonds concentration in the Al powder and density of current will effect on size, porosity and roughness of the coating. UDD particles increase both internal stresses and microhardness in all studied cases. In fact, visual roughness (fig. 4, 6) of the alumina-based coating hardened by rounded diamonds as shown in fig. 1 (b, c) is less then that of alumina hardened with sharp edged particles in fig. 3, 5 at all regimes. In the same way, the diamonds of sharp shape increase porosity. For example, rounded diamonds with 0.15% concentration in the Al powder result in 1.3 nm average grain size, $0.119*10^8$ N/m² internal micro-stresses and 12.7 relative units of texture degree.

3. Observations of technological process show that size of the clusters and its distribution in the coating depends on UDD concentration in the Al substrate, diamonds shape and current density, whereas higher current density increase both the grain size and the internal stresses of the coating. We expect that UDD particles do not effect on electrolyze conditions in view of viscosity, conductivity, pH, and electrode polarization, but it might effects on crystallization mechanism of the coatings. Moreover, the UDD particles could change normal growth of the alumina layer while oxidizing.

Of interest is the process that could be described as follows. Some researches revealed that the process comprises the following stages: anodizing, sparking, micro arc oxidizing and oxidizing. Since the stages of anodizing and sparking are relatively short, the hard oxide structure of the coating is mainly formed at the stage of micro arc oxidizing. At this stage, the processing produces oxide film of 10-15 μ m thickness as it is found to be in traditional anodizing. After that, the process of micro arc oxidizing creates segments of dense alumina matrix, which are growing into the substrate usually in the depth up to 300 μm. Time of processing and growth rate of coating depends on technological regimes and ranges up to two hours for thick layer.

Microplasmic processes are being produced in anodic semi period of alternating current. The observations of the process illustrated that spark of a micro arc appearing during electrical throwing of oxide film takes place along with the process of micro arc oxidizing. Throwing power significantly effects on time of sparking in anodic semi period of alternating current. By increasing the voltage frequency in anodic-cathode regime of the process, the throwing power decreased. In result, an effective part of anodic semi period is increased due to decreasing of pause between the impulses without applied current that is actually in charge for producing the layer and transformations of the Al substrate into alumina.

The diagrams of current and voltage in work piece show that alternating current curve pulse periodically in the sinusoidal form. An analysis of diagrams of the current and voltage applied to the work piece revealed that time of processing at 50 Hz of current frequency decrease the throwing power. In result, the curve of voltage in anodic semi period became smoother via X-axis. The above observations are in good agreement with results reported in [11-14]. In this case, the similar changes in the diagram can be observed. The process passes through all stages in short time at the frequency above 500 Hz in contrast to 50 Hz used in traditional anodizing. The peaks of sinusoidal curve became smoother at the current frequency above 100 Hz. The frequency above 500 Hz intensifies the process of micro arc oxidizing and its structure as well.

0.25 μm

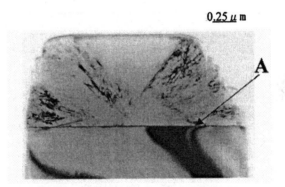

Figure 7. Diamond segment of the composite

Figure 7 shows a high-resolution electron micrograph of diamond grain of the composite. The incident beam was parallel to the <111> axis, with the plane predominantly parallel to the interface. The heavily irradiated region was observed at 10 μm from the surface. High strain and dislocations were generated in the 10-35 μm region of the composite depending on the alumina thickness. However, no difference of lattice constants was distinguished in the Fourier transform (FT) images obtained from the near surface region, damaged region, and deeper-lying region, because the lattice parameter value was determined to be constant in all regions. This assured that no new clusters were in the composite structure.

Based upon the researches, the samples with higher roughness and worst microstructure were not used to study their friction behavior. Figure 8 depicts the effect of diamonds particles on the friction behavior of the alumina-based composite against a stainless-steel ball in the ball-on-disk experiments. Single alumina layer (line 2 in fig. 8) showed high friction force and respectively high coefficient of friction through the sliding distance and increasing applied load. On the contrary, the friction force of alumina-diamonds composite (line 1 in fig. 8) was constant along the measured distance and load ranges. The harder substrate improves frictional behavior of the composite bearing applied load and localized stresses that might induce failure and cracks in alumina coating. The ultimate load for the alumina-diamonds samples was about 90N when friction force increased significantly.

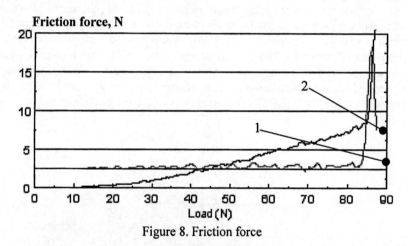

Figure 8. Friction force

XPS scanning analysis was conducted for selected worn surfaces on the composite. The substance on the unhardened wear track was mainly composed of alumina particles of different phases that are mainly gamma phases, which scratched the worn surface of the steel in thickness. On the other hand, hardened alumina exhibits low peeled off abrasive particles on the contacted surfaces that provides better frictional behavior and low abrasive wear of the opposite surface.

CONCLUSION

Ultra dispersed diamonds effects on roughness, microstructure and hardness of the alumina-based composites. Ultra dispersed diamond particles are involved into the polycrystalline alumina-based matrix during micro arc oxidizing of Al-diamonds sprayed layer. Microstructure of the composite became fine with presence of UDD particles and roughness decreases up to Ra 0.5-0.3 μ m. UDD crystals are distributed into alumina-based matrix.

Diamonds concentration in the Al powder and density of current will effect on size, porosity and roughness of the coating. UDD particles increase both internal stresses and microhardness in all studied cases. UDD particles could change normal growth of the alumina layer while micro arc oxidizing of the base substrate.

Observations of technological process show that size of the clusters and its distribution in the coating depends on UDD concentration in the Al substrate, diamonds shape and current density, whereas higher current density increase both the grain size and the internal stresses of the coating. The process passes through all stages in short time at the frequency above 500 Hz in contrast to 50 Hz used in traditional anodizing. The frequency above 500 Hz intensifies the process of micro arc oxidizing and its structure as well.

On the contrary, the friction force of alumina-diamonds composite was constant along the measured distance and load ranges. The ultimate load for the alumina-diamonds samples was about 90 N when friction force increased significantly.

ACKNOWLEDGEMENTS
This work was performed under the scientific direction and principal supervision of Professors Zinovii P. Shulman (Institute of Heat and Mass-transfer, National Academy of Sciences of Belarus) and Prof. P.A.Vitiaz (INDMASH NAS of Belarus).

REFERENCES
[1]Jerry L. Patel, Nannaji Saka, Ph.D. A new Coating Process for Aluminum. Microplasmic Corp., Peabody, Mass. www.microplasmic.com, (2000) [assessed 20 March 2001]
[2]Apticote Ceramic Coatings. A T Poeton & Son Limited, Gloucester, UK. www.poeton.co.uk. [assessed 20 March 2001]
[3]K. Euh, W. S. Kim, K. Shin, S.-H. Lee. Vacuum Compo-Casting Process of Aluminium Alloy Composites Reinforced with SiC. http://www.alumtrans.com. [assessed 20 March 2001]
[4]M. Manoharan, M. Gupta. On Predicting the Fracture Strain of an Age Hardened SiC Reinforced AA6061 Aluminium Metal - Matrix Composite. http://www.alumtrans.com
[5]Berestnev O.V., Basenuk V.L., Kireytsev M.V. et al. Composite coating and technology of its manufacturing. C 23 C 28/00. Accepted patent RU №2000111046/02, 2001.
[6]Khmyl, A.A., Dostanko, A.P., Lanin, V.L., Emelyanov, V.A.: Materials and Technology of Contact Joints Made Using Ultrasonic Microwelding, Proc. of the Sixteenth International Conf. on Electrical Contacts. Loughborough, England, 1992, pp.255-260.

[7]Holden, C.A., Law, H.H., Crane, G.R. Effect of surface X-ray analysis of the galvanic plating. Izvestiya Akademii Nauk BSSR. ser. phys.tech. nauk, №1 (1982), pp. 120-121.

[8]V. Basenuk ed. in chief., M. Kireytsev et al. "Development of fundamentals for technology and design of composite powder materials based on aluminum with nonmetallic components...". Report under the theme "Material-65". Registration Number № 19962944. INDMASH NAS of Belarus, Minsk, (2000), p.165.

[9]Maxim V. Kireytsev. Mechanical properties of composite coatings based on hard anodic oxide ceramics. Proc. of the SEM Annual Conference and Exposition. Best Student Paper Competition. June 4-6, 2001 Portland Marriott Downtown, Portland, Oregon, USA.

[10]William W. Corcoran & Leonid M. Lerner. Hard Anodizing of 2xxx Series Aluminum Alloys. Aluminum Anodizers Council. Southern Aluminum Finishing Company http://www.saf.com/aac.html. [assessed 20 March 2001]

[11]Malishev V.N. Aspects of coating formation by anodic-cathode micro arc oxidizing. // Metal protection. V. 32, №6, pp. 662-662, in Russian, (1991)

[12]Timoshenko A.V. Magurova J.V. Microplasmic oxidizing of Al-Cu aluminum alloys. // Metal protection. V. 31, №5, pp.523-531, in Russian, (1992)

[13]Bakovets V.V. et al. Plasma-electrolytic anodic processing of metals. Novosibirsk, Nauka, Siberian Dept. of NAS of Russia, pp. 168. (1991)

[14]Markov G.A. Micro arc oxidizing. /Vestnik Moscow St. Univ. "Mashinostroenie" №1, in Russian (1992)

Mechanical Behavior

CRACK PROPAGATION AND FRACTURE RESISTANCE BEHAVIOR UNDER FATIGUE LOADING OF A CERAMIC MATRIX COMPOSITE

D.Ghosh and R.N.Singh
Department of Materials Science and Engineering
University of Cincinnati
P.O Box 210012, Cincinnati OH 45221 – 0012

ABSTRACT
The crack growth behavior of unidirectional silicon carbide (SCS-6) fiber reinforced zircon ($ZrSiO_4$) matrix composites under cyclic fatigue is studied in the temperature range 20 –1400°C. In situ crack length measurements under fatigue loading are obtained from single edge notched specimen using a traveling optical microscope. Fracture resistance curves (R-curve) are obtained from the experimental crack growth data, which show a rising R-curve behavior with a decreasing value of the fracture resistance at elevated temperatures. Numerical analyses/iterations, based on the micromechanical models, are done to determine the relationship between the bridging stress function and the R - curve behavior of the composite. The effect of residual stress and constituent properties such as fiber strength, interfacial shear stress, and matrix toughness on the bridging stress function is systematically studied both at room and elevated temperatures.

INTRODUCTION
Ceramic materials have excellent properties such as high strength to weight ratio, high elastic modulus, high thermal conductivity, good oxidation and corrosion resistance and low coefficient of thermal expansion. These properties make them ideal for use as structural components at room and elevated temperatures. However their application is severely limited because of low fracture toughness and inherently brittle nature. CFCCs (Continuous Fiber Ceramic Composites) have the potential for much greater resistance to catastrophic failure than their monolithic counterparts while retaining the excellent properties of ceramics[1-2]. This is mainly due to the presence of the bridging fibers, which remain intact even after the matrix has cracked.[3]

Since CFCCs are intended for high temperature applications, it is of primary importance to understand their high temperature crack propagation and fracture behaviors. A common way for evaluating the toughening effect in ceramic materials is to measure the fracture resistance curve (R- curve). The fracture resistance of a material as a function of the crack growth is referred to as the fracture resistance curve or R – curve.[4-5] The R - curve is a continuous record of toughness development in terms of K_R (stress intensity factor) plotted against crack extension in the material as the crack is driven under a continuously increased stress intensity factor. The objectives of the research work are: (a) study of the in situ crack growth in composites under fatigue conditions both at room and elevated temperatures, (b) determination of the fracture resistance under fatigue loading of unidirectional SiC (silicon carbide) fiber reinforced $ZrSiO_4$ (zircon) matrix composite in the temperature range 20°C- 1400°C, (c) comparison of the experimental R - curve with the theoretical R- curve obtained from numerical analysis and (d) study of the effect of different constituent parameters on the bridging stress function.

EXPERIMENTAL PROCEDURE

For the composites used in the experiments, SCS-6 silicon carbide fibers (Textron Specialty Materials, Lowell, MA) were used as continuous reinforcement materials. The composite was fabricated by uniaxially aligning SiC fibers into a perform and then incorporating the zircon matrix by tape casting and lamination techniques. This was followed by binder burnout in vacuum and in air. The green body was then densified to full density by a hot pressing operation in a nitrogen atmosphere at 1640°C under an uniaxial pressure of 10 MPa. The final fiber volume fraction in this composite was about 25 %.

Fatigue tests were performed on notched composites in three-point flexure mode using a servo hydraulic machine (MTS System Co.) between room temperature and 1400°C in a flowing high purity argon atmosphere. Tests were done under the load control mode at a frequency of 1 Hz. Polished composite samples of dimensions 5 mm x 1.3 mm x 30 mm were tested with an outer span of 20 mm. A notch length of about 1.75 mm (an initial a/W ratio of 0.35) was created using a fine diamond saw in an Ultraslice™ 6000 Precision machine. During the test, a traveling optical microscope (Syncro Vision™, Control Vision Inc.) with a resolution of about 2.5μm was used along with a pulsed light source to measure the crack length. This was done so that an in situ relation between the applied load and crack length could be obtained. The minimum applied load was about 25 N for the first fatigue cycle and the waveform was haversine at all temperatures. For each cycle, the difference between the minimum and the maximum load was 20 N. When crack growth arrest occurred (usually after 1000 cycles) the load was subsequently increased by another 10 N and the sample was subject to another set of fatigue cycles. Hence for each loading cycle the R-value, where R is the ratio of the minimum to the maximum load, was different.

The R- curve was determined from the applied stress intensity factor, K, using the following ASTM standard equation[6]: -

$$K_{app} = \frac{3PS}{2BW^2} \sqrt{a} \ F\ (a\ /\ W) \qquad (1)$$

where

$$F(a/W) = \frac{1.99 - \frac{a}{W}(1-\frac{a}{W})[2.15 - 3.93\frac{a}{W} + 2.7(\frac{a}{W})^2]}{(1+2\frac{a}{W})(1-\frac{a}{W})^{3/2}}$$

where S, B and W are the span, thickness, and width, respectively, of the specimen, a is the crack length, and F (a/W) is a dimensionless function which depends on the crack length a. By measuring the applied load P and the corresponding crack length a, the stress intensity factor K_{app} can be calculated.

RESULTS AND DISCUSSION

Mechanical Behavior of Composites

The typical stress – displacement plot of the composite at room temperature under three point bending is shown in Fig.1. From the figure it is evident that the composites have high toughness (represented by the area under the curve) and strength. The plot shows an initial elastic portion followed by a sudden load drop at the point of first matrix cracking (FMC). Beyond this point, there is an extended regime of increasing load-bearing capacity since most of the reinforcing fibers are still intact and can carry additional loads. The ultimate tensile strength (UTS) represents the maximum load carrying capacity and after this point a gradual drop is observed as more and more of the SiC fibers begin to break and pullout from the matrix. Typical values of FMC and UTS are 340 MPa and 700 MPa, respectively, at room temperature for the samples. Figure 2 shows the load displacement plot for the notched composite samples during fatigue loading for the entire temperature range (20 - 1400°C). The plot shows that the toughness (indicated by the area under the curve) and the ultimate strength decreased with increasing temperature. However, the composite still showed tough behavior, even at 1400°C.

Crack Propagation under Fatigue Loading

Figures 3(a-b) show the nature of crack propagation in the notched composite samples at room temperature and 1400°C. These micrographs were taken after the fatigue tests using an optical microscope. We observe that the crack propagation is different at room and elevated temperatures. At 20°C, 1200°C and 1300°C, a single crack originated from the notch and propagated with increasing load and number of cycles till it reached an a/W ratio of 0.7 beyond which crack deflection and branching occurred. At elevated temperature of 1400°C multiple crack growth at few places in the sample was observed. The crack path at 1400°C is tortuous in contrast to the straight crack path at lower temperatures. The broken fibers in the crack wake are readily visible at 1400°C. The crack opening ahead of the notch also increases with increasing temperature as evident from the micrographs.

Figure 4 shows the crack length vs. number of cycles for the composite at different temperatures. Fatigue experiments were done for about 4000 cycles at a constant value of R (where R is the ratio of the minimum to the maximum load) at each temperature. At room temperature the minimum and maximum loads were 115 N and 135 N (R = 0.85), respectively. The K_R values ranges from 11 - 18 MPa√m. At 1200°C the load range was 105 - 125 N (R = 0.84) and the K_R values ranged from 5 – 9 MPa√m. For the 1300°C tests the corresponding values for the load were 70 - 90 N (R = 0.77) and the K_R range was from 4 – 7 MPa√m. At 1400°C the load ranged from 50 - 70 N (R = 0.71) and the K_R range was from 3 – 6 MPa√m. This is in contrast to the fatigue experiments for the fracture resistance tests where the value of R was changed (since load was increased after each set of cycles). It can be seen that the crack arrest occurred after approximately 1000 cycles, because of the fiber bridging in the crack wake approached a steady state.

Fracture Resistance Behavior (R - curve) under Fatigue Loading
Figure 5 shows the stress intensity at crack initiation at various temperatures. The K_R value at crack initiation decreases with increasing temperature. At room temperature this value is 10.8 MPa√m compared to 3.67 MPa√m at 1400°C. At high temperatures the fiber strength decreases and the bridging zone is not well developed and hence crack grows at lower stress intensity values.

The experimental data of fracture resistance (R - curve) obtained from in situ crack growth and Eq. (1) are shown in Fig. 6. The data show an initial fracture toughness of 3 – 10 MPa√m and then a continuous increase with increasing normalized crack length a/W at each temperature. Crack deflection and fiber matrix debonding occurs in front of crack tip and fiber bridging develops in the crack wake and hence we see a rising R - curve with increasing crack length. The slope of the crack resistance curves decreased with increasing temperature, and this reduction in slope is very significant at 1400°C. Therefore, we can conclude that resistance to crack propagation decreases significantly above ~1300°C.

Theoretical Fracture Resistance Curve

The theoretical calculations of the K_R at different temperatures were done using a numerical approach developed by Llorca et al. and Singh et al.[7-9] in which both debonding and bridging of intact fibers and fiber pull–out processes were considered. Computer programs based on numerical iterative technique were used to calculate the theoretical R- curve by using the room and elevated temperature mechanical properties of the SCS-6 fiber, zircon, and SCS-6 fiber/zircon matrix composites. The basic equation describing the contribution of the bridging stress to the stress intensity factor is given by:

$$K_R = K_M + \int_{a_o}^{a} \frac{2\sigma_b(u)}{\sqrt{\pi a}} H(\frac{a}{W}, \frac{x}{a}) \, da \qquad (2)$$

where K_M is the matrix toughness, $\sigma_b(u)$ is the bridging stress function which is dependent on the crack opening, $u(a,x)$, a_o is the initial crack length, W is the specimen width, and H is the weight function depending on the specimen geometry and loading condition.

Comparison Between Experimental and Theoretical Results

Figure 7 shows the results of the numerically calculated theoretical K_R curves at room temperature with and without taking into consideration the residual stress. The experimental data for the fracture resistance curves from the in situ optical method are also plotted along with the best-fit curve for the experimental data. The plot was drawn using the following values, $\sigma_f = 2.8$ GPa, $m=3$, and $\tau = 6.3$ MPa.[8, 10] We observe that the theoretical R-curve determined without taking residual stress into consideration is higher than that determined with the residual stress at all ranges of a/W.

Figure 8 shows the results of the experimental and numerical K_R curves at 1400°C. The values of the constituent properties used for this plot are as follows: $\sigma_f = 0.8$ GPa, $m=3$, and $\tau = 1$ MPa. From the plot it is evident that the theoretical K_R curves generated with the value of $K_m = 7$ MPa$\sqrt{m}$ and $\tau = 1$ MPa provided higher values than the experimental data at 1400°C. These results also indicate that the numerical data does not match well with the experimental data at all a/W values at elevated temperature. A plausible reason for this behavior might be the onset of significant creep deformation of the fiber and matrix phases above 1300°C, and the weakening of the fibers due to creep.[12]

Fiber Bridging Stress in the Composites

The bridging stress function can generally be modeled by doing an analysis of the stress transfer between the bridging fibers and the matrix, as well as the associated specimen geometry and loading conditions. Hence, the experimental R-curve can be associated to $\sigma_b(u)$ by applying the basic bridging–stress law and specimen dependent weight function. The equations involved and the micromechanical models used for the analysis are described by equation (2), and reference [8].

Figure 9 show the influence of changes in fiber strength, Weibull modulus, interfacial shear strength, and residual stress of fiber on the bridging stress $\sigma_b(u)$-crack opening (u) displacement behaviors at room temperature. These curves exhibit an initial rising portion controlled by the debonding and bridging processes of the intact fibers, and decreasing portion dominated by the fiber pullout and fiber fracture processes.[8] Figure 9 shows the curves for the case when the residual stress is taken into account as well as for a case of zero residual stress. An initial bridging stress of (–45) MPa is determined from the graph, which decreases as the crack opening increases.

Figure 10 shows the bridging stress $\sigma_b(u)$ vs. crack opening displacement (u) behavior at 1400°C. The parameters (σ_f =0.8 GPa, m=3, τ=1 MPa) used for the solid line curve are the same as used to calculate the best-fit R-curve at 1400°C. Fiber strength is varied from 0.8 GPa to 1.3 GPa, Weibull modulus values are varied from 3 to 5 while the interfacial shear stress values varied from 1 to 3 MPa. However, the maximum values of the bridging stress and areas under the curve have dramatically decreased at 1400°C. This is consistent with the low value of the fracture resistance at 1400°C.

CONCLUSIONS

The crack propagation and fracture resistance behaviors under fatigue loading of silicon carbide fiber/ zircon matrix composites were studied both experimentally and numerically in the temperature range 20°-1400°C by a novel in situ method. Tough ceramic composites showed a rising R – curve at all temperatures with the maximum value of the stress intensity decreasing with increasing temperature. The effect of temperature was to shift the R - curves to progressively lower K_R values. The crack opening at the notch tip increased with increasing temperature. At 20°, 1200°, and 1300°C a straight crack propagated from the notch while the crack path was tortuous at 1400°C. The crack propagation rate gradually decreased with increasing number of cycles until crack arrest occurred after approximately 1000 cycles at all temperatures.

A numerical approach was used to calculate the theoretical R-curve at all temperatures. At 20°, 1200°, and 1300°C, the theoretical curve fitted reasonably well with the experimental curve but at 1400°C there was a significant difference between the numerical and experimental curves. Residual stress in the fiber affected the theoretical R- curve at room temperature whereas influence of the residual stress of fiber on the numerically calculated R-curves was negligible at temperatures ≥ 1200°C.

Fiber strength (σ_f), Weibull modulus (m), and interfacial shear strength (τ) affected the bridging stress, $\sigma_b(u)$ vs. crack opening displacement (u) curves. Of these parameters, fiber strength was found to have the strongest influence on the bridging stress function. The maximum values of the bridging stress and the area under the curve progressively decreased as the temperature was increased. This was consistent with the lower values of the fracture resistance observed at elevated temperatures.

ACKNOWLEDGEMENT

This material is based on upon work supported by National Science Foundation Under Grant No. 0070213 . Any opinions, findings, and conclusions or recommendations expressed in this material are those of the authors and do not necessarily reflect the views of the National Science Foundation.

REFERENCES

[1] A. G. Evans, "Perspective on the Development of High –Toughness Ceramics," *Journal of the American Ceramic Society*, **73**, [2], 187- 206 (1990).

[2] R. Raj, " Fundamental Research in Structural Ceramics for Service Near 2000° C," *Journal of the American Ceramic Society*, **76**, [9], 2147- 2174 (1993).

[3] M.D.Thouless and A.G.Evans, " Effects of Pull Out on the Mechanical Properties of Ceramic Matrix Composites," *Acta Metallurgica et Materiala*, **36**, [11], 517-522 (1988).

[4] R.W. Steinbrech, A. Reichel, and W. Schaarwacher, " R – Curve Behavior of Long Cracks in Alumina," *Journal of the American Ceramic Society,* **73**, [7], 2009-2015 (1990).

[5] J. Llorca and R. N. Singh, "Influence of Fiber and Interfacial properties on Fracture Behavior of Fiber-Reinforced Ceramic Composites," *Journal of the American Ceramic Society,* **74**, [11], 2882-90 (1991).

[6] Standard Test Method for Determination of Fracture Toughness of Advanced Ceramics at Ambient Temperature, *ASTM Standard*, C1421-99 (2000).

[7] J. Llorca, M.Elices, and J.A. Celemin, " Toughness and Microstructural Degradation at High Temperatures in SiC Fiber Reinforced Ceramics," *Acta Metallurgica et Materiala*, **46**, [7], 2441-2453 (1998).

[8] Y.L.Wang, U. Anandakaumar, R.N. Singh, "Effect of Fiber Bridging Stress on the Fracture Resistance of Silicon – Carbide Fiber/ Zircon Composites," *Journal of theAmerican Ceramic Society,* **83**, [5], 1207-1214 (2000).

[9] J. Llorca and M. Elices, "A Simplified Model to Study Fracture Behavior in Cohesive Materials," *Cement and Concrete Research*, **20**, 92- 102 (1990).

[10] R.N. Singh, "Interfacial Properties and High Temperature Mechanical Behavior of Fiber – Reinforced Ceramic Composites," *Materials Science and Engineering A*, **166**,185-198 (1993).

[11] B. Budiansky, J. W. Hutchinson and A. G. Evans, "Matrix Fracture in Fiber-Reinforced Ceramics," *Journal of the Mechanics and Physics of Solids*, **34** [2], 167-189 (1986).

[12] U. Anandakumar, "Matrix Cracking and Creep Behavior of Monolithic Zircon and Zircon Silicon Carbide Fiber Composites," PhD. Dissertation, University of Cincinnati, Cincinnati, OH, (2000).

[13] A. Sayir and S. Farmer, Personal Communication, NASA Glenn Research Center.

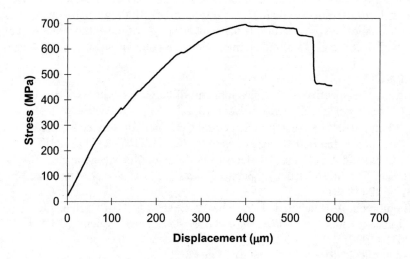

Figure 1. Stress displacement curve for a SiC_f / zircon matrix composite at room temperature tested in three point bending .

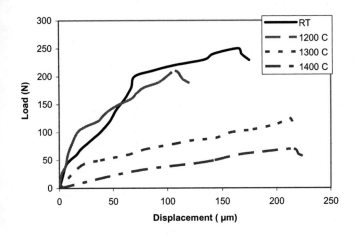

Figure 2. Load displacement curves for the notched composite at various temperatures.

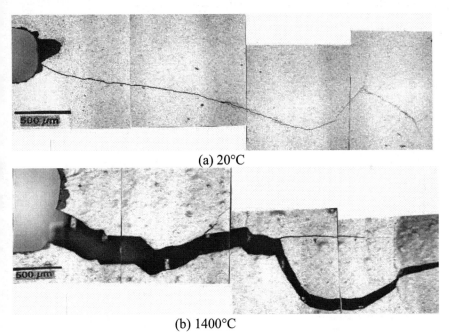

(a) 20°C

(b) 1400°C

Figure 3. Crack propagation in the notched specimen at (a) 20°C and (b) 1400°C.

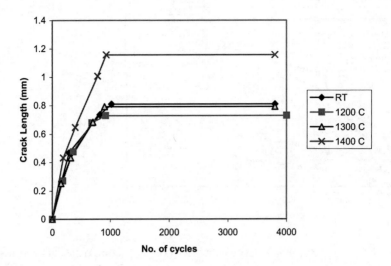

Figure 4. Variation of crack length with number of cycles at various temperatures.

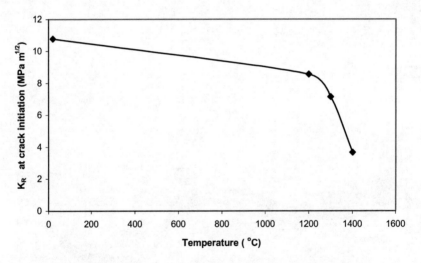

Figure 5. Variation of K_R (at crack initiation) with temperature.

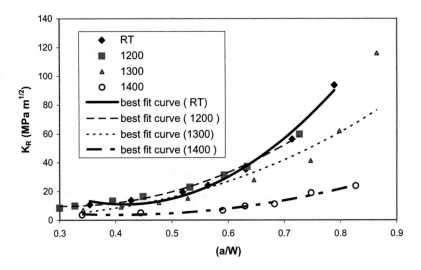

Figure 6. Experimental fracture resistance curve (R- curve) at different temperatures.

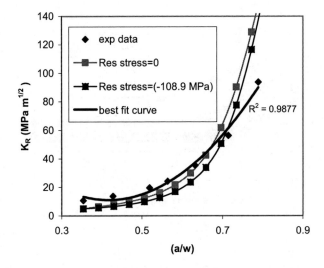

Figure 7. A comparison of the experimental and theoretical R–curve behaviors in zircon–SiC$_f$ composites at room temperature. The influence of residual stress in fibers on the theoretical curves is also illustrated.

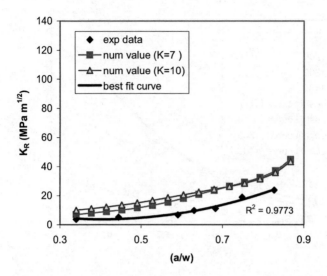

Figure 8. Experimental and theoretical R-curves based on numerical calculations at 1400° C.

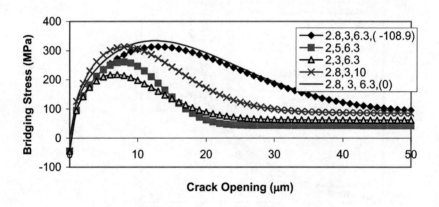

Figure 9. The effect of residual stress on the bridging stress functions generated using numerical calculations at room temperature. The curve without the residual stress is also shown in this plot. The values inside the bracket in the legend box are the residual stresses.

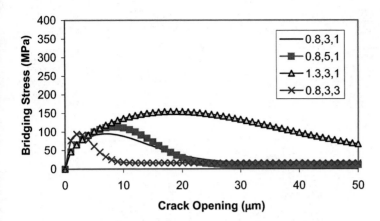

Figure 10. Results of the bridging stress functions generated using numerical calculations at 1400° C. The three values (for each set of data) in the legend box represent the fiber strength (σ_f), Weibull modulus (m), and the interfacial strength (τ), respectively.

SPECIMEN SIZE EFFECT ON THE IN-PLANE SHEAR PROPERTIES OF SiC/SiC COMPOSITES

Takashi Nozawa, Yutai Katoh and Akira Kohyama
Institute of Advanced Energy, Kyoto University
Gokasho, Uji, Kyoto 611-0011, Japan

Edgar Lara-Curzio
Metals and Ceramics Division, Oak Ridge National Laboratory
Oak Ridge, TN37831-6069

ABSTRACT

Miniaturization of test specimens is often necessary to evaluate the physical and mechanical properties of materials under severe environments. The validation of these techniques requires an understanding of the role of geometric and volumetric (size) effects on the mechanical behavior. However, little work has been focused on the effect of these variables on shear properties, on comparison to tensile properties. This paper presents the results of a study aimed at assessing the effect of notch separation and specimen thickness on the shear strength of a 2-D SiC/SiC composite by the Iosipescu test method. It was found that in-plane shear strength had a specimen geometric dependency with no relation to the specimen gauge size. This dependency was quite opposite to that on the off-axis tensile strength. This indicated that in-plane shear was not the critical fracture factor in the off-axis tension. Other mechanism such as the detachment strength at the fiber and matrix (F/M) interface might make a significant effect on the composite strength.

INTRODUCTION

Small specimen test technique (SSTT) for tensile testing has been developed to meet several demands in advanced energy industries [1, 2]. First of all, in the field of fusion research, SSTT has been focused on one of the effective means to evaluate irradiated materials, because of the achievement to reduce radiological hazard potential. SSTT also has an advantage to make the distribution of defects uniform in the specimen body as well as possible. This is very important to simplify the evaluation of composite characteristics with the complex fracture

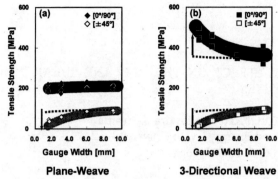

Fig. 1 Width effect on tensile strength of (a) P/W and (b) 3-D SiC/SiC composites [10]

behavior. In addition, statistical analysis will be available as the number of specimens increasing. And what is the most important is to make it possible to evaluate mechanical properties of small components for the practical applications. And needless to say, this technique has long been strongly required to establish, because of the maximum use of the lab-scale products with very limited volumes.

However, each component has the finite size and so this makes it difficult to minimize specimen size because of the specimen size effect [3, 4]. The weakest link theory is often utilized to discuss it [5]. This says that quasi-brittle materials such as monolithic ceramics lose its strength as the specimen volume increasing. This is because unavoidable defects in composites, the weakest one of which becomes a critical origin for the composite failure, exist more in larger volume. Size effect is also closely dependent on the fracture mode, fabric architecture, loading direction, environment and others [6, 7].Therefore, it is very important to identify the key factors to govern the size effect, which will be applicable to the estimation of the composite strength in any size.

Tensile properties of [0°/90°] SiC/SiC composites had a size effect, dependent on the architecture, according to our previous research [8, 9]. Composite strength was significantly dependent on the properties of the reinforcing fiber and they directly affect on it. Tensile strength of 3-D SiC/SiC was determined by the fiber strength itself and hence it had a gauge length effect that is explained by the weakest link theory. Moreover, it is well known experimentally and theoretically that the maximum strength was roughly proportional to the volume fraction of fibers aligned in the tensile axis [5, 8, 9]. However, the fiber volume fraction for the 3-D configuration was different in shorter gauge width due to the structural limitation. When the gauge width is shorter than the width of the structural minimum unit, some specimens have many

Advances in Ceramic Matrix Composites VIII

fibers and others have much matrix with the relation to the location of cutting. Hence, tensile properties, which had a close relation to fiber properties, had also specific size dependency related to the axial fiber volume fraction as shown in Fig. 1. On the contrary, 2-D SiC/SiC, whose fibers were tightly weaved, had less size dependency.

However, tensile properties in the off-axis tension had quite different size dependency from that of [0°/90°] SiC/SiC. It is well known that the composite strength significantly decreased because of the anisotropy due to its characteristic architecture, once the loading direction changed from the longitudinal direction of fibers [10]. This is because the change of the fracture mode from tension of fibers to shear or detachment at the fiber/matrix interface might cause to the reduction of the composite strength. And what is the most important is that the change of fracture mode made the size dependency on tensile properties quite different. In fact, tensile strength decreased as the gauge width decreasing for both architectures, once the loading direction came apart from the longitudinal direction of fibers [10]. This is due to the change of the fracture mode from tension of fiber to shear strength between bundles or debonding strength at F/M interface, as mentioned before. Then it will be more important to identify the size dependency in each mode for the complete understanding.

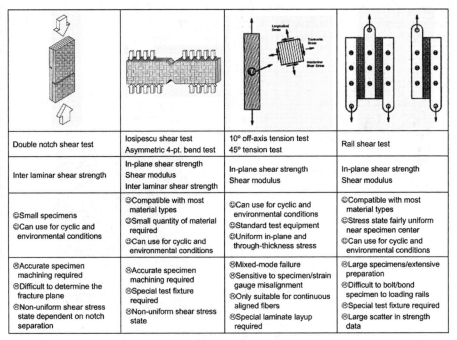

Double notch shear test	Iosipescu shear test Asymmetric 4-pt. bend test	10° off-axis tension test 45° tension test	Rail shear test
Inter laminar shear strength	In-plane shear strength Shear modulus Inter laminar shear strength	In-plane shear strength Shear modulus	In-plane shear strength Shear modulus
☺Small specimens ☺Can use for cyclic and environmental conditions	☺Compatible with most material types ☺Small quantity of material required ☺Can use for cyclic and environmental conditions	☺Can use for cyclic and environmental conditions ☺Standard test equipment ☺Uniform in-plane and through-thickness stress	☺Compatible with most material types ☺Stress state fairly uniform near specimen center ☺Can use for cyclic and environmental conditions
☹Accurate specimen machining required ☹Difficult to determine the fracture plane ☹Non-uniform shear stress state dependent on notch separation	☹Accurate specimen machining required ☹Special test fixture required ☹Non-uniform shear stress state	☹Mixed-mode failure ☹Sensitive to specimen/strain gauge misalignment ☹Only suitable for continuous aligned fibers ☹Special laminate layup required	☹Large specimens/extensive preparation ☹Difficult to bolt/bond specimen to loading rails ☹Special test fixture required ☹Large scatter in strength data

Fig. 2 Shear test methodologies for ceramic matrix composites

Shear strength is distinguished into more details by the correlation between the directions of the working force and aligned fibers. In-plane shear strength and inter-laminar shear strength (ILSS) are well known ones. The former works equally in each plane of stacked sheets, which are perpendicular to stacking direction. The latter works in the stacking direction of fabric sheets and delaminates them. Short beam test and double notch shear test are usually used for the evaluation of ILSS, and rail shear test, asymmetric four-point flexural test, Iosipescu shear test, and off-axis tensile test are often used for the measurement of in-plane shear strength and shear modulus (Fig. 2) [11]. Off-axis tensile test method has an advantage to obtain precise stress-strain relationships, however, at the same time, this has a difficulty in the evaluation of pure shear stress due to the co-existence of shear and tensile failure modes [12]. On the contrary, Iosipescu shear test and asymmetric four-point flexural test enabled to form only the shear stress field between two V-notches and this makes it useful to evaluate in-plane shear strength. However, extensive stress concentration at the contact with fixture often caused some cracks critical for the composite failure.

The objectives in this study are to investigate size dependency of in-plane shear strength, which was considered to make a significant influence on the off-axial tensile properties, and, in particular, to distinguish the volumetric effect and geometric effect on the shear properties. In further investigation, relationships between tensile and shear properties were evaluated. Also, Iosipescu test methodology for the evaluation of the in-plane shear was discussed in more details.

EXPERIMENTAL

A plane-weave (P/W) SiC/SiC composite was prepared by Ube Industries, Ltd., Japan. This composite had stitching fibers in the transthick direction. First of all, Si-Ti-C-O fiber (Tyranno™-LoxM, Ube Industries, Ltd., Japan) used as

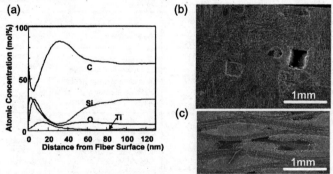

Fig. 3 As-received material; (a) compositional profile at the fiber surface, (b) surface image and (c) cross-sectional image.

reinforcements had undergone a surface modification before weaving in order to optimize the fiber-matrix interface. Due to this treatment, compositionally-graded interphase with excess carbon was formed close to the fiber surfaces (Fig. 3) [13]. Afterwards, these composites were synthesized by polymer impregnation and pyrolysis (PIP) process. This composite had relatively high porosity above 10 % after the several PIP sequences and especially inter-bundle porosities near stitching fibers were characteristic. From this reason, the bulk density was quite lower, about 2.2 Mg/m^3.

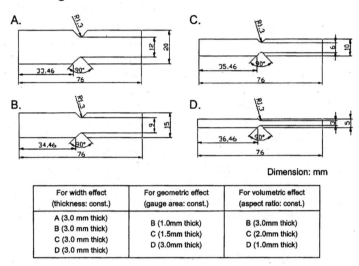

Dimension: mm

For width effect (thickness: const.)	For geometric effect (gauge area: const.)	For volumetric effect (aspect ratio: const.)
A (3.0 mm thick) B (3.0 mm thick) C (3.0 mm thick) D (3.0 mm thick)	B (1.0mm thick) C (1.5mm thick) D (3.0mm thick)	B (3.0mm thick) C (2.0mm thick) D (1.0mm thick)

Fig. 4 Schematic illustrations of Iosipescu shear specimens for size effect researches

Fig. 5 Schematic illustration of test fixture

In order to evaluate size effect on the in-plane shear strength, Iosipescu shear test was conducted by using hydraulic mechanical testing machine (MTS 809 Axial/Torsional Test System). Schematic illustrations of test specimens and the test fixture were as shown in Fig. 4 and 5. Distance between notches and the specimen thickness were varied in each specimen. Specimens for the evaluation of the volumetric effect were designed as possessing the same aspect ratio, i.e. notch separation to thickness, however different volume, respectively. On the contrary, specimens for the evaluation of the geometric effect had the same volume but different aspect ratio. All the tests were carried out on the guideline of ASTM C1292. Crosshead speed was at 0.6 mm/min. Shear stress was calculated as load divided by the effective area, which means the fracture area measured after testing. After the tests, fracture surfaces were characterized by the optical microscopy and the scanning electron microscopy (SEM).

RESULTS AND DISCUSSION
Iosipescu Shear Fracture Behavior
Typical stress-displacement relation in Iosipescu tests was shown in Fig. 6. All the curves obtained in this study had an excellent response, in particular linear relation in the beginning until proportional limit. Just beyond proportional limit, it became non-linear due to the accumulation of matrix cracks. At the maximum, large stress drops occurred due to the failures of 90° directionally aligned fibers by tensile stress worked in the transverse-loading (90°) direction. At this time, visible large cracks emerged between two V-notches. Beyond this point, the friction of pullout fibers maintained all the applied loads. After the short term of gradual decreases of loads, stress dropped again. This means complete pullouts of broken fibers. Beyond this point, no tensile and frictional forces were worked.

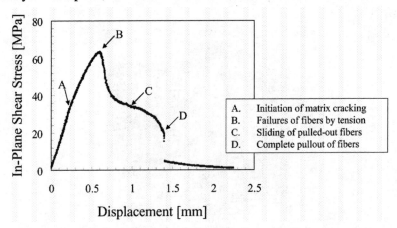

Fig. 6 Typical stress-displacement response of Iosipescu shear test

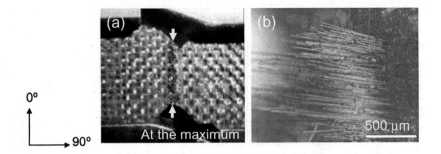

0°

90°

Fig. 7 Typical fracture images of Iosipescu shear test; (a) crack propagation between two notches and (b) long fiber pullouts

According to the fracture surface, main crack propagated between fiber bundles, not in intra bundles (Fig. 7). Pulled-out fibers were very numerous. These fracture appearances were very similar to those obtained in tension. This is because most fibers were broken by tension worked in the 90° direction.

Volumetric Effect and Geometric Effect on Shear Properties

Fig. 8 shows the size dependency of the maximum shear stress, i.e. in-plane shear strength. This figure shows a gradual increase of the in-plane shear strength in the smaller gauge width. In order for the further discussion about the size dependency, the volumetric effect and the geometric effect on shear properties were investigated (Fig. 9). It was clearly shown that strength increases for the specimen with the smaller aspect ratio, i.e. notch separation to thickness, was characteristic, in case of the constant volume in the gauge section. With increasing

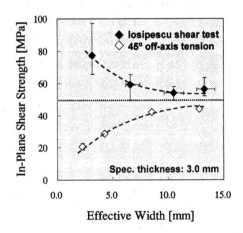

Effective Width [mm]

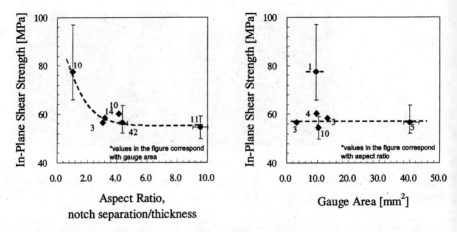

Fig. 9 Volumetric effect and geometric effect on in-plane shear strength

the aspect ratio, shear strength tended to converge into some constant value (Fig. 9 (a)). On the contrary, in case of the constant aspect ratio, there was no clear difference in the specimen size (Fig. 9 (b)). This means that the size dependency of the in-plane shear strength was considered due to the geometric effect. In conclusion, this implies that composites with the small aspect ratio have a strength increase regardless of the specimen size. However, it is necessary to investigate the mechanism of the strength increase for the complete understanding.

Relationships between In-Plane Shear and Tension

Fig. 8 shows the width dependency of the tensile strength on comparison to that of the shear property. This figure points out some reasons for the reduction of the off-axis tensile strength in the short gauge width. The size dependency of the in-plane shear strength showed an opposite tendency to that of the off-axis tension. This is an apparent contradiction. This fact indicates that the key fracture mechanism was quite different in the off-axis tension, at least in the shorter gauge width. In general, three kinds of stresses; tension in the longitudinal fiber direction, transverse tensile stress perpendicular to the longitudinal fiber direction, and in-plane shear parallel to fibers, are working in the off-axis tension. In the shorter gauge width, tensile strength in the off-axis tension might be determined by the weak detachment strength at the fiber/matrix interface not but the fiber tension and the in-plane shear, since there was no complex cross of weaving and each fiber was easy to detach with no special restriction by the weaving architecture. On the contrary in the larger width, composite strength might be determined by the mixture of them, since fiber bundles were complexly weaved. In short, the

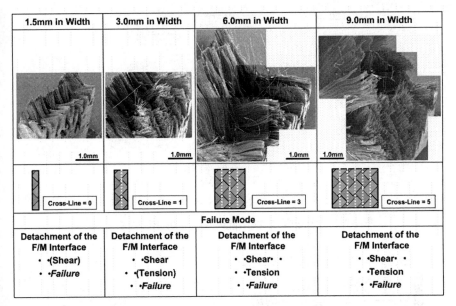

1.5mm in Width	3.0mm in Width	6.0mm in Width	9.0mm in Width
1.0mm	1.0mm	1.0mm	1.0mm
Cross-Line = 0	Cross-Line = 1	Cross-Line = 3	Cross-Line = 5
Failure Mode			
Detachment of the F/M Interface • •(Shear) • •Failure	Detachment of the F/M Interface • •Shear • •(Tension) • •Failure	Detachment of the F/M Interface • •Shear• • • •Tension • •Failure	Detachment of the F/M Interface • •Shear• • • •Tension • •Failure

Fig. 10 Fracture behaviors of 45° off-axis tension of SiC/SiC composites

fracture mode was differed with the gauge width in the off-axis tension, because of the existence of several failure modes simultaneously (Fig. 10).

Evaluation of Iosipescu Shear Test Methodology

Iosipescu shear test, which requires a specific fixture, is one of the effective means for the evaluation of the in-plane shear strength, because of its simplicity and the formation of pure in-plane shear field between two V-notches. Indeed, most tests (over 92 %) in this study were completed with a high-accuracy. However, it requires some specific cautions in order for the use of small specimens, different from those in ASTM standard.

In case of using support tabs due to the limitation of the fixture size, the in-plane shear strength differed with the location of the support tabs. When tabs were located far from notches, strength loss was significant, especially for specimens with the shorter notch separation. Then, most of them failed at the root of support tabs (Fig. 11). This is due to the large stress concentration at the end of tabs and also the presence of bending moment. Finite element method analyses supported this fact and the non-uniform stress field was produced near the notches and end tabs. This kind of stress concentration was significant in case of the shorter notch separation (Fig. 12). In order to obtain the pure in-plane shear field and analyze it precisely, mounting the specimen is one of the most critical issues,

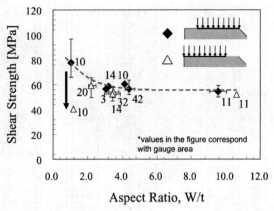

Fig. 11 Strength reduction due to support tab effect

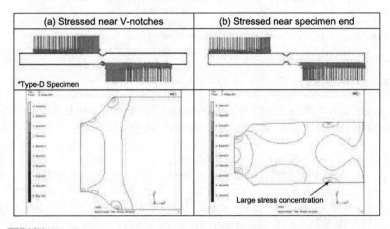

Fig. 12 Stress distributions obtained by FEM analyses

in addition to the high-accuracy of the specimen machining.

CONCLUSIONS

In order to determine the size effect of the in-plane shear strength of SiC/SiC composites and to discuss the relationship between tensile and shear properties, Iosipescu test was performed. Key conclusions are summarized as follows.

1. In-plane shear strength by Iosipescu test showed a specimen geometric dependency. Maximum shear stress was nearly constant with no relation to

the specimen gauge area. However, it tended to increase as the aspect ratio, notch separation to thickness, decreasing.

2. Size effect on the in-plane shear strength was quite opposite to that on the off-axis tensile strength. This indicated that the in-plane shear was not the critical fracture factor in the off-axis tension. Other mechanism such as the detachment strength at the F/M interface might make an effect on the stress reduction in small sized material. Further investigation about this is strongly required.

3. Iosipescu shear test is one of the effective test methodologies for the evaluation of the in-plane shear strength. However, this test requires a uniform stress field between two notches without stress concentration at support tabs and bending portion.

ACKNOWLEDGMENTS

The authors would like to express their sincere appreciation to Dr. M. Sato (Ube Industries Ltd., Japan) for providing the materials. This work is performed as a part of 'R&D of Composite Materials for Advanced Energy Systems' research project supported by Core Research for Evolutional Science and Technology (CREST) and supported by collaboration at Japan-US Program for Irradiation Test of Fusion Materials (JUPITER-II).

REFERENCES

[1]T. Nozawa, K. Hironaka, Y. Katoh, A. Kohyama and E. Lara-Curzio, "Small Specimen Test Technique for Tensile Testing of SiC/SiC Composites," *Jurnal of Nuclear Materials*, to be published.

[2]Y. Kohno, "The Evaluation of Mechanical Properties by Means of Small Specimen Test Technology," *Journal of Plasma and Fusion Research*, **76-4** 368-375 (2000).

[3]L.S. Sutherland, R.A. Shenoi and S.M. Lewis, "Size and Scale Effects in Composites: I. Literature Review," *Composites Science and Technology*, **59** 209-220 (1999).

[4]L.S. Sutherland, R.A. Shenoi and S.M. Lewis, "Size and Scale Effects in Composites: II. Unidirectional Laminates," *Composites Science and Technology*, **59** 221-233 (1999).

[5]W.A. Curtin, "Theory of Mechanical Properties of Ceramic-Matrix Composites," *Journal of the American Ceramics Society*, **74** 2837-2845 (1991).

[6]M. Ibnabdeljalil and W.A. Curtin, "Strength and Reliability of Fiber-Reinforced Composites: Localized Load-Sharing and Associated Size Effects," *International Journal of Solids and Structures*, **34** 2649-2668 (1997).

[7]M.R. Wisnom, "Size Effects in the Testing of Fibre-Compostie Materials," *Composites Science and Technology*, **59** 1937-1957 (1999).

[8]T. Nozawa, T. Hinoki, Y. Katoh, A. Kohyama, E. Lara-Curzio and M. Sato, "Influence of Specimen Geometry on Tensile Properties of 3D SiC/SiC Composites", *Advances in Ceramic Composites VI, Ceramic Transactions*, **124** 351-362 (2001).

[9]T. Nozawa, T. Hinoki, Y. Katoh, A. Kohyama and E. Lara-Curzio, "Specimen Size Effects on Tensile Properties of 2-D/3-D SiC/SiC Composites", *ASTM STP 1418*, to be published.

[10]T. Nozawa, Y. Katoh, A. Kohyama and E. Lara-Curzio, "Effect of Specimen Size and Fiber Orientation on the Tensile Properties of SiC/SiC Composites," *Ceramic Engineering and Science Proceedings*, to be published.

[11]W.R. Broughton, "Shear"; pp. 100-123 in *Mechanical Testing of Advanced Fibre Composites*, Edited by J.M. Hodgkinson, Woodhead Publishing Ltd., Cambridge England, 2000.

[12]L. Denk, H. Hatta, A. Misawa and S. Somiya, "Shear Fracture of C/C Composites with Variable Stacking Sequence," *Carbon*, **39** 1505-1513 (2001).

[13]T. Ishikawa, K. Bansaku, N. Watanabe, Y. Nomura, M. Shibuya and T. Hirokawa, "Experimental Stress/Strain Behavior of SiC-Matrix Composites Reinforced with Si-Ti-C-O Fibers and Estimation of Matrix Elastic Modulus," *Composites Science and Technology*, **58** 51-63 (1998).

SOLID-PARTICLE EROSION OF ZrSiO4 FIBROUS MONOLITHS

K. C. Goretta, F. Gutierrez-Mora, T. Tran, J. Katz, J. L. Routbort
Argonne National Laboratory, Argonne, IL 60439-4838, USA

T. S. Orlova
A. F. Ioffe Physico-Technical Institute, 194021 St. Petersburg, Russia

and A. R. de Arellano-López
University of Sevilla, 48010 Sevilla, Spain

ABSTRACT

Erosive damage was studied in ZrSiO4 monolithic and fibrous-monolithic specimens subjected to impact at 90° by angular SiC particles traveling at 50 or 70 m/s. Steady-state erosion rates in the fibrous monoliths were higher than would be predicted by a rule of mixtures based on erosion rates of the cell and cell-boundary phases. The relatively rapid erosion was attributed to easy removal of the cell boundary, followed by larger-scale loss of individual cells that had lost support because of the removal of the surrounding cell boundary.

INTRODUCTION

Ceramic fibrous monoliths (FMs) consist of a strong cellular phase surrounded by a weaker cell-boundary phase that effectively dissipates energy during fracture. FMs are produced in-situ from powders and thus have inherent cost advantages over fiber-reinforced ceramic composites. Moreover, FMs fail gracefully in flexure and can support significant loads to large strains [1-9].

FMs have been produced from many materials systems, including nitrides, carbides, borides, and oxides. Among commercial products are Si3N4/BN, SiC/graphite, various carbides and borides, and cermets [10]. The best combination of room-temperature mechanical properties to date (fracture strengths to ≈700 MPa and work-of-fracture values to ≈9 kJ/m2) has been obtained from FMs in which BN is the cell-boundary phase [2-9]. These FMs also offer promise for excellent performance at elevated temperature, although an effective coating will be required for long-term service in oxidizing environments. Oxide FMs may be more suited to certain applications in oxidizing environments.

Oxide FMs have generally contained porous cell boundaries to promote debonding and crack deflection during fracture [11-14]. Recently developed porous-matrix oxide-fiber composites and multilayers exhibit excellent mechanical properties [15,16], and the expectation is that analogous FMs can be fabricated. FMs produced from oxides such as Al_2O_3 and $ZrSiO_4$ have proved moderately successful: they fail gracefully, but generally at stresses of <150 MPa [11-14].

Erosion of Si_3N_4/BN FMs has been studied [17]. Erosion rates were higher in the FMs than would be predicted by a rule of mixtures [18,19] based on the erosion rates of the two constituents. However, the strengths of monolithic Si_3N_4 and BN specimens were substantially reduced by erosion, but the strengths of the FMs were independent of erosive damage [17]. $ZrSiO_4$ FMs are toughened by a different mechanism from that of Si_3N_4/BN FMs. The response of porous-matrix FMs to solid-particle erosion has been unknown, and whether their strengths will also prove to be independent of severe surface damage has been an open question. This work was undertaken to assess the response of a porous-matrix FM to solid-particle erosion.

EXPERIMENTAL METHODS

FMs were fabricated from two $ZrSiO_4$ powders by a sequential method. In the first step, duplex monofilaments consisting of a core of <1 μm $ZrSiO_4$ Alfa Aesar powder (Ward Hill, MA) and a sheath of >10 μm Remet $ZrSiO_4$ powder (Utica, NY) were coextruded. Twin extrusion screws were fed into a single die, inside which the sheath was formed around the core. In the second step, filaments were cut and bundled and then ram-extruded to produce multifilaments. In the final forming step, these filaments were cut, laid up by hand, and cold-pressed to produce green unidirectional FM bars. After burnoff of the organics, the FM specimens were sintered in air at 1550°C for 3 h. Details of the processing procedures and formulations have been published [11-13].

The FM bars experienced ≈23 vol.% shrinkage between initial pressing and the fired state. They were ≈65 vol.% cell and ≈35 vol.% cell boundary, and the average cell size was ≈150 μm (Fig. 1). This structure differs significantly from the ≈80-85 vol.% cell and ≈15-20 vol.% cell boundary generally observed for Si_3N_4/BN FMs [2-9]. The rather high fraction of cell boundary could be attributed, at least in part, to the large particle size of the Remet $ZrSiO_4$ powder.

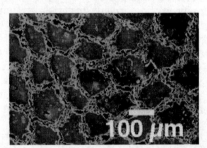

Fig. 1. SEM photomicrograph of transverse
cross section of typical $ZrSiO_4$-based FM.

Monolithic bars of Alfa Aesar and Remet ZrSiO$_4$ were also prepared. Powders of each were cold-pressed at 70 MPa and the resulting bars were fired in air at 1550°C for 3 h. The Alpha Aesar bars were ≈95% dense; the Remet bars were ≈70% dense [13]. Erosion tests were conducted in an evacuated slinger-type apparatus [20,21]. The erodent was 143-μm-diameter angular SiC particles (Norton, Worcester, MA). The impact angle was 90° and the impact velocity, V, was 50 or 70 m/s. Steady-state erosion rates, ER, expressed in mg/g, were obtained as mass lost (ΔW) from each specimen per mass of SiC particles impacting the specimen. From three to five cycles of testing were conducted to define each steady-state value. Specimens were weighed before each cycle, eroded, cleaned with a brush, and then reweighed. Scanning electron microscopy (SEM) was conducted on all eroded specimens.

RESULTS AND DISCUSSION

Weight-loss data were obtained for one FM specimen, two dense specimens (representative of the cells), and one porous specimen (representative of the cell boundary). For each set of experiments, the linear least-squares fit to the weight-loss data was excellent, and reproducibility between the two dense specimens was within 2% (Fig. 2). ER values obtained at the two impact velocities are shown in Fig. 3. Only two velocities were used because we had insufficient FM material for additional testing.

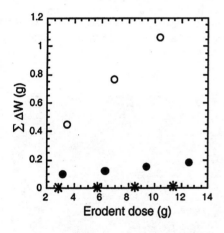 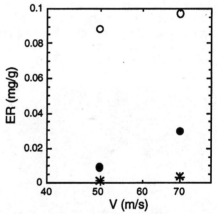

Fig. 2. Accumulative weight loss vs. erodent dose at V = 50 m/s: Remet ZrSiO$_4$ (open circles), Alpha ZrSiO$_4$ (+ and X), and FMs (filled circles).

Fig. 3. Erosion rate vs. impact velocity for erosion of various specimens: Remet ZrSiO$_4$ (open circles), Alpha ZrSiO$_4$ (+ and X), and FMs (filled circles).

The erosion data for each specimen appeared to be generally consistent with prevailing models for erosion of brittle materials. In each model,

$$ER \propto V^n, \tag{1}$$

where, depending on the assumption of impact conditions, n ranges from 2.0 to 3.2 [22-24]. The data were taken for only two velocities over a relatively small range; detailed analysis of the value of n is therefore unwarranted. However, the value obtained for the FM was ≈ 3.5, which is higher than expected and higher than observed for the monolithic materials. This relatively strong dependence on the energy of impact may be related to the mechanism of mass loss. This speculation is addressed in the discussion of the SEM observations of the eroded specimens.

For composites such as the ZrSiO4 FM, it has been shown that erosion rates should be related to the rates of the individual constituents by

$$1/ER_{FM} = m_C/ER_C + m_{CB}/ER_{CB}, \tag{2}$$

where m refers to the mass fraction of the constituent in the FM, and the subscripts FM, C, and CB refer to the FM, cell, and cell-boundary specimens, respectively [18]. This model holds so long as the erosion rates of the constituents in the FM are independent of each other. Comparison of erosion rates indicates that the FM was eroded approximately five to six times faster than would be predicted by this rule of mixtures. SEM observations revealed the cause of this anomalously fast erosion.

SEM was first conducted on the eroded monolithic specimens. The denser, finer-grained ZrSiO4, composed of Alfa Aesar powder, was eroded in a manner typical of brittle, polycrystalline ceramics. Its eroded surfaces were rough and there was significant evidence of cleavage (Fig. 4).

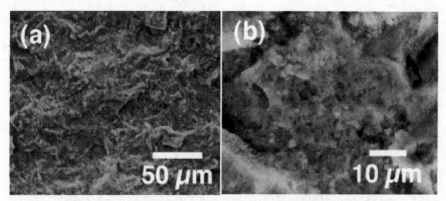

Fig. 4. SEM photomicrographs of eroded surfaces of denser monolithic ZrSrO4; small, well-distributed porosity and brittle fracture are observed in both (a) and (b).

The porous and coarser-grained $ZrSiO_4$, composed of Remet powder, appeared to be eroded primarily by wholesale removal of grains (Fig. 5a). Its eroded surfaces were rough and granular in appearance. There was minor evidence of cleavage of individual grains (Fig. 5b).

The surface of the eroded FM resembled in many places that of the Alfa Aesar specimen and in many places that of the Remet specimen. Oblique views clearly indicated more rapid erosion of the porous cell boundary (Fig. 6). Many cells had clearly been fractured and removed in large segments, rather than having been eroded by a series of small-scale damage events (Fig. 7). This wholesale removal of cells was doubtless responsible for the relatively high erosion rates of the FM and was related to the strong velocity dependence of the FM erosion rate.

Similar high erosion rates relative to predictions of the applicable rule of mixtures have also been measured for Si_3N_4/BN FMs [17]. For both those FMs and the $ZrSiO_4$ FMs reported on here, the average cross-sectional dimension of the cells was approximately equal to that of the impacting particles. Under such erosive conditions, it appears to be a general finding that preferential removal of the less-erosion-resistant cell boundary leads to fracturing and removal of poorly supported cells. Anomalously high erosion rates result.

Fig. 5. SEM photomicrographs of eroded surfaces of highly porous monolithic $ZrSiO_4$: weight loss appears to have been due primarily to removal of individual grains (a), but some fracturing of grains was evident (b).

SUMMARY

$ZrSiO_4$ fibrous monoliths and monolithic specimens of their two constituents were subjected to impact at 90° by angular 143-μm-diameter SiC particles traveling at 50 or 70 m/s. Erosion occurred by brittle fracture in the denser monolithic $ZrSiO_4$ and by grain removal and fracture in the porous $ZrSiO_4$. Steady-state erosion rates in the FMs were higher than would be predicted by a rule of mixtures based on erosion rates

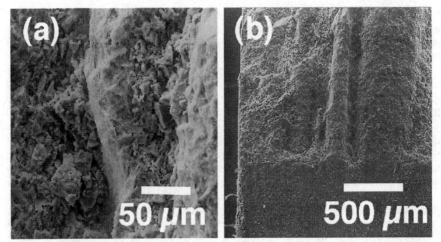

Fig. 6. SEM photomicrographs of eroded surfaces of ZrSiO₄ FM: (a) surface containing large regions similar to that observed for eroded Remet specimens; (b) preferential removal of cell boundary.

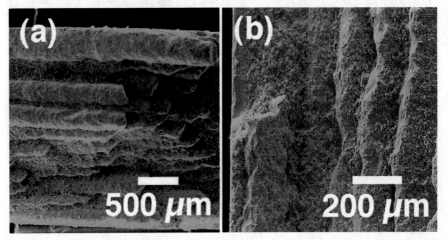

Fig. 7. SEM photomicrographs of eroded surfaces of ZrSiO₄ FMs; fracture and removal of large segments of cells are observed in (a) and (b).

of the cell and cell-boundary phases. The relatively rapid erosion was attributed to removal of the cell boundary, followed in many places by removal of large segments of individual cells that had lost support because of loss of surrounding cell boundary.

ACKNOWLEDGMENTS
We thank our colleagues J. J. Picciolo and B. J. Polzin for help with specimen preparation. This work was supported by the Defense Advanced Research Projects Agency through a U.S. Department of Energy Interagency Agreement, under Contract W-31-109-Eng-38, and by North Atlantic Treaty Organization Grant PST.CLG.977016.

REFERENCES
1. W. S. Coblenz, "Fibrous monolithic ceramic and method for production," U.S. Patent 4,772,524, Sept. 20, 1988.
2. D. Popovic', J. W. Halloran, G. E. Hilmas, G. A. Brady, S. Somas, A. Bard, and G. Zywicki, "Process for preparing textured ceramic composites," U.S. Patent 5,645,781, July 8, 1997.
3. G. A. Danko, G. E. Hilmas, J. Halloran, and B. King, "Fabrication and properties of quasi-isotropic silicon nitride-boron nitride fibrous monoliths," *Ceram. Eng. Sci. Proc.* **18** [3], 607-13 (1997).
4. G. Hilmas, A. Brady, U. Abdali, G. Zywicki, and J. Halloran, "Fibrous monoliths: non-brittle fracture from powder-processed ceramics," *Mater. Sci. Eng.* **A195**, 263-8 (1995).
5. D. Kovar, B. H. King, R. W. Trice, and J. W. Halloran, "Fibrous monolithic ceramics," *J. Am. Ceram. Soc.* **80**, 2471-87 (1997).
6. S. W. Lee and D. K. Kim, "High-temperature characteristics of Si_3N_4/BN fibrous monolithic ceramics," *Ceram Eng. Sci. Proc.* **18** [4], 481-6 (1997).
7. R. W. Trice and J. W. Halloran, "Influence of microstructure and temperature on the interfacial fracture energy of silicon nitride/boron nitride fibrous monolithic ceramics," *J. Am. Ceram. Soc.* **82**, 2502-08 (1999).
8. M. Tlustochowitz, D. Singh, W. A. Ellingson, K. C. Goretta, M. Rigali, and M. Sutaria, "Mechanical-property characterization of multidirectional Si_3N_4/BN fibrous monoliths," *Ceram. Trans.* **103**, 245-54 (2000).
9. J. C. McNulty, M. R. Begley, and F. W. Zok, "In-plane fracture resistance of cross-ply fibrous monoliths," *J. Am. Ceram. Soc.* **84**, 367-75 (2001).
10. Product brochure, Advanced Ceramics Research, Tucson, AZ (2001).
11. B. J. Polzin, T. A. Cruse, R. L. Houston, J. J. Picciolo, D. Singh, and K. C. Goretta, "Fabrication and characterization of oxide fibrous monoliths produced by coextrusion," *Ceram. Trans.* **103**, 237-44 (2000).
12. A. J. Mercer, G. E. Hilmas, T. A. Cruse, B. J. Polzin, and K. C. Goretta, "A comparison study of the processing methods and properties for zirconium silicate fibrous monoliths," *Ceram. Eng. Sci. Proc.* **21** [3], 605-12 (2000).

13. K. C. Goretta, et al., "Development of Advanced Fibrous Monoliths: Final Report For Project of 1998-2000," ANL-01/04 (Argonne National Laboratory Report, 2001).

14. S.-J. Lee and W. M. Kriven, "Toughened oxide composites based on porous alumina-platelet interphases," *J. Am. Ceram. Soc.* **84**, 767-74 (2001).

15. C. G. Levi, J. Y. Yang, B. J. Dalgleish, F. W. Zok, and A. G. Evans, "Processing and performance of an all-oxide ceramic composite," *J. Am. Ceram. Soc.* **81**, 2077-86 (1998).

16. W. J. Clegg, "Design of ceramic laminates for structural applications," *Mater. Sci. Technol.* **14**, 483-95 (1998).

17. K. C. Goretta and J. L. Routbort, Argonne National Laboratory, unpublished information (2002).

18. S. K. Hovis, J. E. Talia, and R. O. Scattergood, "Erosion in multiphase systems," *Wear* **108**, 139-55 (1986).

19. J. L. Routbort and R. O. Scattergood, "Solid particle erosion of ceramics and ceramic composites," *Key Eng. Mater.* **71**, 23-50 (1992).

20. J. L. Routbort, D. A. Helberg, and K. C. Goretta, "Erosion of whisker-reinforced ceramics," *J. Hard Mater.* **1**, 123-35 (1990).

21. P. Strzepa, E. J. Zamirowski, J. B. Kupperman, K. C. Goretta, and J. L. Routbort, "Indentation, erosion, and strength degradation of silicon-alloyed pyrolytic carbon," *J. Mater. Sci.* **28**, 5917-21 (1993).

22. A. G. Evans, M. E. Gulden, and M. E. Rosenblatt, "Impact damage in brittle materials in the elastic-plastic response regime," *Proc. Roy. Soc. London Ser. A* **361**, 343-65 (1978).

23. S. M. Wiederhorn and B. R. Lawn, "Strength degradation of glass impacted with sharp particles: I, annealed surfaces," *J. Am. Ceram. Soc.* **62**, 66-70 (1979).

24. J. E. Ritter, P. Strzepa, K. Jakus, L. Rosenfeld, and K. J. Buckman, "Erosion damage in glass and alumina," *J. Am. Ceram. Soc.* **67**, 769-74 (1984).

Characterization

NONDESTRUCTIVE EVALUTATION (NDE) AND TENSION BEHAVIOR OF NEXTEL/BLACKLAS COMPOSITES

J. Kim, B. Yang, and P. K. Liaw
Department of Materials Science and Engineering, the University of Tennessee
Knoxville, TN 37996-2200

H. Wang
Metals and Ceramics Division, Oak Ridge National Laboratory
Oak Ridge, TN 37831

ABSTRACT

Tension behavior of Nextel™ 312 fiber reinforced Blackglas™ ceramic matrix composites (CMCs) was investigated with the aid of nondestructive evaluation (NDE) techniques. Several NDE methods, such as ultrasonic testing (UT), infrared (IR) thermography, and acoustic emission (AE) techniques, were employed for this investigation. Prior to tensile testing, UT was used to characterize the initial defect distribution of Blackglas™ CMC samples, i.e., developing ultrasonic C-scans. During tensile testing, AE sensors and an IR camera were used for in-situ monitoring of progressive damages of CMC samples. AE provided the amounts of damage evolution in terms of the AE intensity and/or energy, and the IR camera was used to obtain the temperature changes during the test. UT was conducted on fractured samples after tensile testing to compare progressive damages with the initial defects. Microstructural characterization using scanning electron microscopy (SEM) was performed to investigate fracture mechanisms of Nextel™/Blackglas™ samples. In this paper, NDE techniques were used to facilitate a better understanding of fracture mechanisms of Nextel™/Blackglas™ during tensile testing.

INTRODUCTION AND OBJECTIVES

Nextel™/Blackglas™ ceramic matrix composites (CMCs) have been developed as a potential candidate for aerospace and military usages, especially for high-temperature structural applications, such as advanced propulstion systems, rocket motors, and space vehicle protection, due to their better thermal and mechanical shock resistance, excellent corrosion and wear resistance, lightweight, and a good balance of toughness and strength at high temperatures [1]. Nextel™/Blackglas™ composites are relatively new material, and the composites were developed by the low-cost ceramic composites (LC3) project [2]. However, not many mechanical properties data are available so far.

Nondestructive Evaluation (NDE) techniques, such as ultrasonic testing (UT), infrared (IR) imaging, and acoustic emission (AE), can be powerful methods to investigate the fracture behavior and defect distribution in the composite [3-4]. However, relatively little work has been performed on relating the NDE signatures to the understanding of mechanical behavior of CMCs.

The main objectives of this investigation are to introduce nondestructive evaluation (NDE) techniques for assuring the quality and structural integrity of Nextel™/Blackglas™ composites, to perform NDE using UT, IR thermography, and AE methods for the analyses of defect distributions that may affect mechanical properties, to investigate tension behavior of Nextel™/Blackglas™ composites with the aid of NDE techniques, and finally, to provide fracture and NDE information to aid in the fabrication, development, and selection of Nextel™/Blackglas™ composites for structural applications.

MATERIALS SYSTEM AND EXPERIMENTAL PROCEDURES
Materials System

The used materials for this study is continuous Nextel™ (manufactured by 3M) fiber reinforced Blackglas™ matrix composites, designated as Nextel™/Blackglas™. Nextel™ is an elliptical continuous ceramic fiber with a diameter of about 10 to 12 μm, and the chemical composition of Nextel™ fiber is 62% Al_2O_3, 24% SiO_2, and 14% B_2O_3 in weight percent (wt%). The material was supplied as ceramic prepreg and consisted of a ceramic reinforcement fiber along with a preceramic polymer, Blackglas™. The chemical composition of the Blackglas™ matrix is 42% Si, 20% C, and 38% O_2 in wt%. The fiber volume content is 43.3%. The fiber was woven into a crow foot satin weave style designated AF-30, i.e., satin weaved with 4 harness. Boron nitride (BN) was used for coating material of Nextel™ fibers to provide a weak interface between the Nextel™ fibers and Blackgals™ matrix to induce more fiber pullout during failure. Figure 1 presents the microstructures of Nextel™/Blackglas™ composites.

Ultrasonic Testing (UT)

NDE techniques were used to characterize the defect distribution and to provide the defect and damage evolution information of composites before, during, and after mechanical testing. Nondestructive characterization with as-received specimens could provide inherent defect information and/or distributions, which may affect the fracture behavior during mechanical testing. At first, UT was employed to Blackglas™ composite specimens to provide a two-dimensional defect distribution as a function of ultrasonic transmitted amplitudes.

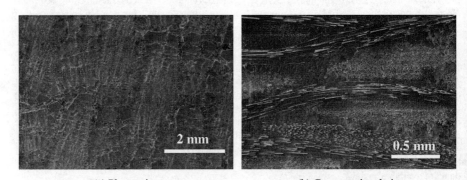

(a) Planar view   (b) Cross-sectional view

Fig. 1. The microstructures of Nextel™/Blackglas™ composites; (a) planar view and (b) cross-sectional view, respectively.

The ultrasonic NDE is the most widely used technique for the detection and characterization of composite materials [3-5]. Ultrasonic testing is a nondestructive method in which beams of high-frequency sound waves are introduced into materials for the detection of both surface and internal flaws in the material. The UT sound waves travel through the material with some attendant loss of energy and are deflected at interfaces and/or defects. The deflected beam can be displayed and analyzed to assess the presence and location of flaws or discontinuities. In this research, ultrasonic amplitude measurements were performed using a pulse-echo C-scan mode at a frequency of 15 MHz, in an immersion tank.

Infrared (IR) Thermography
   Thermography is a powerful NDE technique for the characterization of composite materials. Since the composite materials show relatively high emissivities, composites are suited for examinations with or without surface treatments [3]. The use of an infrared (IR) camera can provide rapid, non-contact scanning of surfaces, components, or assemblies [6-8]. The high-speed IR camera provided the measurement of temperature change during tensile testing as well as the images of failure contour at the time of fracture in this investigation. The characteristics of the IR camera for the current investigation include the maximum speed of 40 Mega-pixels per second with 14 bit resolution. The focal plane arrays show up to 640×512 pixels, the spatial resolution is about 5.4 μm, and the temperature resolution is as sensitive as 0.015°C at 23°C.

Acoustic Emission (AE)
   When a material is placed under stress, it experiences plastic deformation, formation of flaws, or fracture, which produces small stresses or ultrasonic waves in the material, and acoustic emissions are generated. For ceramic materials, an increase in AE events occurs before fracture, providing a potential means of either detecting crack initiation of a component or monitoring when failure is imminent. Acoustic emissions can be detected by AE sensors, and the piezoelectric transducer converts wave pulses that strike into electrical impulses, which are amplified and displayed [9].
   The AE system hardware used in this investigation consists of piezoelectric sensors, preamplifiers, filters, amplifiers, a signal processing computer, a display monitor, and a printer. A two-channel mode was used, and the sensors were adhesively bonded to the surface of the tensile specimens by silicone rubber, which insured good bonding between the sensor and specimen. Data consisting of load, time, and AE parameters (energy, events, hits, counts, amplitude, rise time, counts to peak, and duration) were recorded in the hard drive of the main processor unit during testing. In this study, two-channel AE sensors were used to monitor the progressive damage during mechanical testing. The AE was capable of predicting failures by showing an increase in the energy/count rate prior to failures during tensile testing.

Tensile Testing
   Monotonic tensile tests were conducted using a computer-controlled Material Test System (MTS) 810 servohydraulic frame equipped with hydraulic grips. The edge-loaded specimen geometry and the hydraulic grips provided accurate and reproducible specimen alignment. The tensile tests were conducted at room temperature under displacement control at a cross-head speed of 0.5 mm per minute. In this study, dog-bone type flat specimens were used to investigate the damage evolution during the tests. The IR camera and AE sensors were used for in-situ monitoring of the temperature change and AE events, respectively, during the tests. Aluminum tabs were attached with epoxy glue at each side of the shoulder section of the coupon in order to avoid rupture of the grip parts during loading the sample.

RESULTS AND DISCUSSION
Tensile Testing
   Figure 2 shows the stress-strain curves obtained from monotonic tensile tests for Nextel™/Blackglas™ composites. Totally, 6 tensile tests were performed in this investigation,

and the results show very similar trends of stress-strain behavior from each test. From those data, the average ultimate tensile strength (UTS) and Young's modulus were determined with 87 MPa and about 27.5 GPa, respectively. Figure 3 shows the stress-strain curves of selected samples for further investigation regarding NDE and mechanical performances.

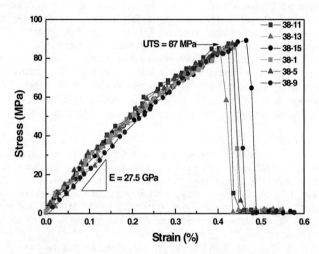

Fig. 2. The stress-strain behaviors of Nextel™/Blackglas™ composites.

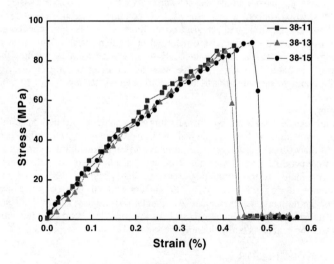

Fig. 3. The close-up of Fig. 2. Three samples were selected for further investigation regarding NDE and mechanical performances.

UT C-scans

Figure 4 shows the results of UT C-scans for Nextel™/Blackglas™ composite specimens before and after mechanical testing. The range of the ultrasonic transmitted amplitude is Min (minimum) to Max (maximum). Min is the smallest relative ultrasonic transmitted amplitude, and Max is the largest. Dark colors stand for high-amplitude signals, generally good or dense regions in the composite, while light colors correspond to low amplitudes, that is, bad or porous regions. Note that the presence of flaws and/or porosity in the traveling path of the ultrasound gives rise to the deflection or energy loss of the traveling ultrasound, and relatively lower values of ultrasonic amplitudes would be obtained. In Fig. 4, a similar trend has been obtained for all three samples. Although some parts containing light colors, showing relatively lower amplitude values, overall pretty good amplitudes values are shown throughout the gage sections of the samples.

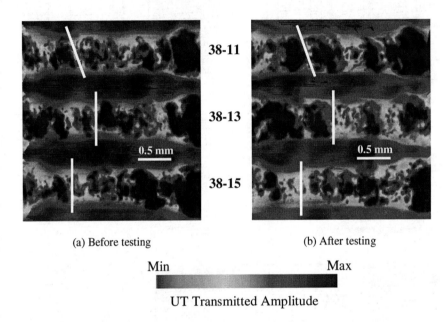

(a) Before testing   (b) After testing

Min   Max

UT Transmitted Amplitude

Fig. 4. UT C-scans for (a) before tensile testing, and (b) after testing. Note that the lines indicate the locations of final failure.

The white lines in Fig. 4 indicate the final fracture positions, i.e., the final ruptures occurred along the white lines. The failures occurred at lower UT transmitted amplitude values. For samples 13 and 15, the failures occurred along 90° directions to the loading direction. However, in the sample 11's case, failure occurred along 30° to the loading direction. Also, in sample 11, two pieces of sample separated completely, while samples 13 and 15 were not separated completely after tensile testing stopped. Therefore, in the sample 11's case, due to the water intake between two separated pieces of sample during the UT C-scan, the UT amplitude values are somewhat higher compared with the amplitude values before testing, which are undesirable results for this investigation. However, for the cases of samples 13 and 15, the damage evolutions in terms of UT transmitted amplitude after tensile testing are shown in Fig. 4. This means that the damage and/or crack propagation areas are showing much lower UT amplitude values than those before testing with light colors. This is mainly due to the presence of macro and micro-cracks, and fiber pullout around fracture surfaces.

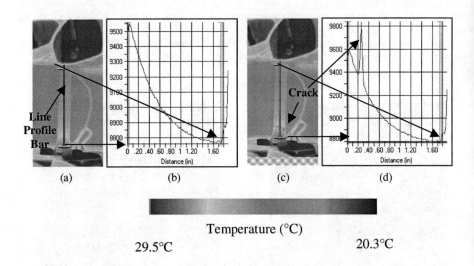

Fig. 5. Temperature evolution during tensile testing for sample 38-11; (a) an IR image before failure, (c) at the time of failure, and (b) and (d) temperature profiles along the gage length for (a) and (c) pictures, respectively.

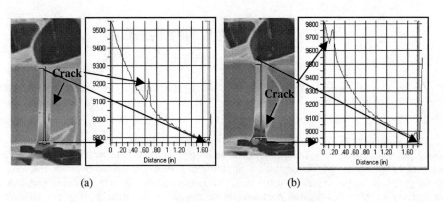

Fig. 6. IR images and temperature profiles for (a) sample 38-13 and (b) sample 38-15, respectively. Note that the temperature scale is the same as shown in Fig. 5.

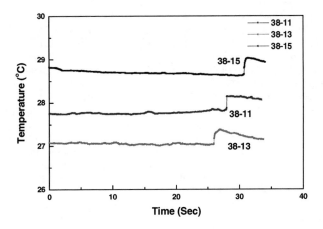

Fig. 7. Temperature changes in terms of time during tensile tests.

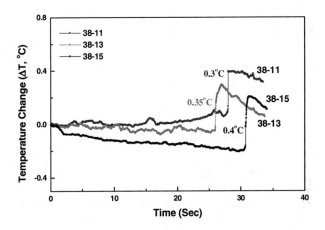

Fig. 8. Temperature peaks found at the time of the failure from tensile testing.

IR Images and Temperature Evolution

Figure 5 shows temperature evolution during tensile testing. The temperature peak has been observed at the time of failure. The speed of the infrared (IR) camera was 60 Hz. Figure 5 also presents the temperature evolution with the IR image and temperature profiles during tensile testing for sample 38-11. The IR images were taken before failure [Fig. 5(a)] and at the time of failure [Fig. 5(c)]. Figures 5(b) and (d) illustrate the temperature profiles along the gage section of the tensile coupon. Especially, the temperature peak has been obtained when the crack is present as shown in Fig. 5(d). Note that the temperature profiles in Fig. 5(c) and (d) were obtained along the length direction of the tensile sample with the marked line profile bar in Figs. 5(a) and (c). The lower ends of the line profile bars indicate starting points (0 inch in distance) of the horizontal axes, as shown in Figs. 5(b) and (d), and the upper ends of the line profile bars in Figs. 5(a) and (c) exhibit the ends of the distance in the horizontal axes as shown in Figs. 5(b) and (d). The vertical axes of Figs. 5(b) and (d) indicate IR unit, i.e., the arbitrary intensity of the temperature. After the calibration of the IR unit, 1,000 IR unit was about 1°K. Since the lower grip section holds the load from the main actuator of the testing machine, the temperature of the lower grip part is somewhat higher than that of the upper grip section. However, the overall temperature change along the length direction of the sample is not significant, only less than 1°C, i.e., not too much temperature change has been observed during tensile testing. Actually, the final crack rupture has been found at the lower part of the specimen, and the temperature peak has been observed. Since the IR camera speed was 60 Hz, 60 frames of pictures were taken every second. After going over the entire frame set, it has been found that only 3 frames were showing the presence of cracks and the temperature peaks around the rupture area. This means that the failure occurred within only 0.05 second. It is possible to infer that failure occurred very brittle and in an instant manner. Also, the temperature change is not so high, only 0.4°C as shown in Figs. 5(b) and (d). Now, we can understand that the failure occurred in a very brittle mode and not much fiber pullouts. Figure 6 presents the IR images and temperature profiles for samples 13 and 15, respectively. The very similar testing results have been obtained, and shown in Fig. 6.

Figure 7 shows temperature changes in terms of time during tensile tests. Overall, not too much temperature changes have been shown, but some temperature peaks have been observed at the time of failure, although the change is only less than 0.4°C, as presented in Fig. 8. It could be inferred that the frictional forces and/or interfacial effects between the fiber and matrix could affect the temperature change during the test. As described in Figs. 5 and 6, the failures were occurred in an instant mode, i.e., the brittle failure mode. In Fig. 8, the temperature evolution behavior of three different samples was somewhat different. The sample 15 shows a temperature decrease until final failure, while both samples 11 and 13 indicate temperature increases before failures. However, the temperature changes of the samples are not too much, only less than 0.2°C from the beginning of the test to before failure. This means that the initial temperature changes are related with the environmental effect of ambient-temperature testing condition, such as temperature fluctuation or the machine noise due to the servohydraulic test frame.

It is difficult to mention that the temperature effects are related to only the initial thermoelastic effect, which generally shows a temperature drop due to the initial elastic behavior of the testing sample. In this investigation, the load level of each sample is relatively low, i.e., the maximum load is about 3.11 kN, therefore, initial minor temperature drop due to the thermoelastic effect could be very small. Also, the small temperature drop because of the thermoelastic effect could be interfered with environmental noise from the hydraulic machine.

Figure 9 presents the relation between the stress and temperature change during tension tests. Figure 9 also shows a direct comparison between the stress-strain behavior and temperature evolution behavior during monotonic tensile testing. As previously described, the final ruptures occurred in an instant mode for all samples. The temperature increases at the time of failure are mainly due to frictional heating between the fiber and matrix during final tensile loading. The mechanisms of frictional heating include debonding at the interface of the fibers and matrix and final fiber pullout. As expected from the stress-strain curves, Nextel™/Blackglas™ composites do not show a graceful failure mode, which is typically found in most ceramic composite materials. In this respect, it is suggested that the failure occurred in a moment within 0.1 second with instant matrix cracking, debonding at the interface, and final fiber pullout. The relatively strong interface of the BN coating also provided very brittle failures. Figure 10 shows another evidence of instant failure in Nextel™/Blackglas™ composites. Figure 10 presents the cumulative acoustic emission

(AE) energy-time relationship during tensile testing, and also, stress-strain curves are juxtaposed with AE results. Only one significant peak has been found during tensile testing, which indicates instant failure of the composite specimen. The main purpose of the use of the AE method for this investigation is to detect and investigate damage initiation and the associated evolution process. Based on this point of view, the failure mode does not show sufficient crack propagation behavior. The failure mode was an instant brittle failure right after crack initiation in Nextel™/Blackglas™ samples.

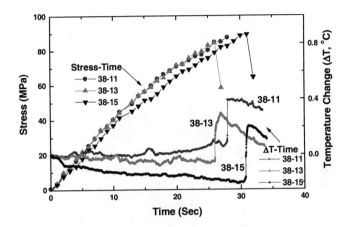

Fig. 9. Stress responses vs. temperature evolution during tensile testing for Nextel™/Blackglas™ composites.

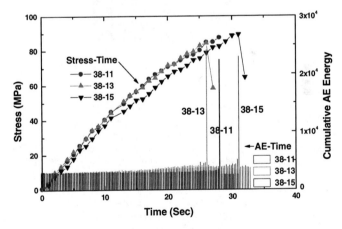

Fig. 10. Stress responses vs. cumulative acoustic emission (AE) energy with respective to time during tensile testing for Nextel™/Blackglas™ composites; AE energies increase as failures are imminent.

Microstructural Characterization using SEM

After mechanical testing, microstructural characterization was conducted to investigate fracture mechanisms in Nextel™/Blackglas™ composites using scanning electron microscopy (SEM). Figure 11 presents the entire cross-sectional view of failure area. The SEM micrograph shows that initial cracks/damages start at 0° fiber bundles and propagated with debonding at the interface between the fibers and matrix. The 90° fiber bundle plays a role to retard the crack propagation, and the bundle provides a toughening mechanism of Nextel™/Blackglas™ composites. Figures 12 and 13 exhibit the evidence of the brittle failure manner, i.e., showing limited fiber pullout (Fig. 12) and strong fiber/matrix interfaces and microcracks in the matrix region and limited fiber pullout in fiber-rich areas (Fig. 13). The brittle Blackglas™ matrix region shows that the microcracks and debris of the matrix material, and the brittle failure mode also observed in fiber-rich area not showing debonding of fiber and matrix, i.e., a relatively strong interface between the Nextel™ fiber and matrix. This phenomenon was also found from previous mechanical testing and NDE results.

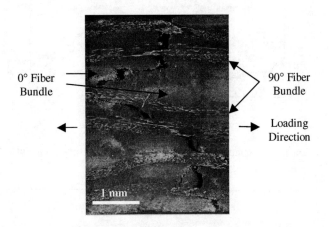

0° Fiber
Bundle

90° Fiber
Bundle

Loading
Direction

1 mm

Fig.11. The entire cross-sectional view of failure area.

Figure 14 presents the summary of the failure mechanisms in Nextel™/Blackglas™ composites showing the fracture surfaces. The initial crack starts from the intersection of 0° and 90° fiber bundles due to the stress concentration, and the crack propagates along the 0° fiber bundle and matrix region (Fig. 11), and the final rupture occurs at 90° fiber bundle with fiber/matrix debonding and fiber pullout. Figure 14 also shows limited fiber pullout.

CONCLUSIONS

The tensile behavior of Nextel™/Blackgals™ composites was investigated with the aid of several NDE techniques, such as ultrasonic testing, acoustic emission, and infrared thermography methods. Various types of NDE methods provided the possible ways to predict and interpret mechanical performances of Nextel™/Blackglas™ composites. The combination of several NDE techniques could assure a greater level of reliability for ceramic composites evaluation to investigate the fracture behavior of Nextel™/Blackglas™ composites. The tensile properties data of Nextel™/Blackglas™ composites were obtained.   The failure mode (mechanisms) of

Nextel™/Blackglas™ was provided through the microstructural characterizaion with SEM micrographs. The fractures of Nextel™/Blackglas™ were shown in brittle failure manners for all testing samples. The relatively strong interfaces between the Nextel™ fibers and Blackglas™ matrices provides brittle fracture modes, not showing a 'graceful period' with extensive fiber pullout.

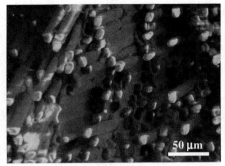

Fig. 12. SEM micrograph showing limited fiber pullout.

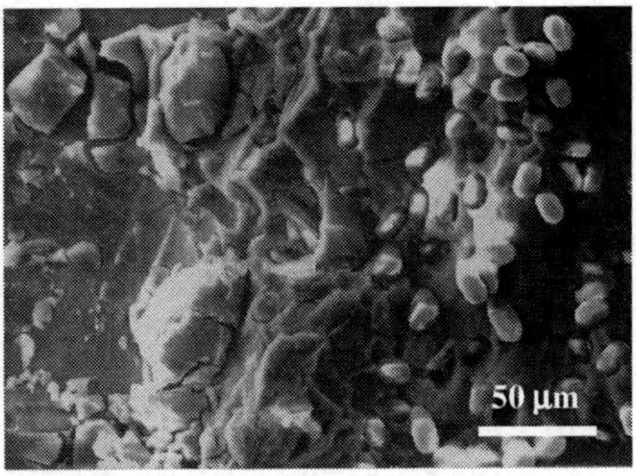

Fig. 13. Brittle failure manner showing strong fiber/matrix interfaces and microcracks in the matrix region.

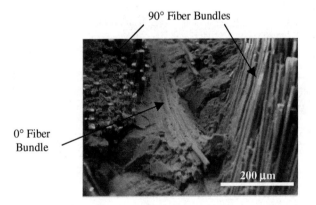

Fig. 14. Fracture surface of Nextel™/Blackglas™ composites. The SEM micrograph was taken from sample 38-15.

ACKNOWLEDGMENTS

This work is supported by the National Science Foundation, the Combined Research Curriculum Development (CRCD) Program, under the contract No. EEC-9527527, the Division of Design, Manufacture, and Industrial Innovation, under DMI-9724476, and the Integrative

Graduate Education and Research Training (IGERT) Program, under DGE-9987548, to the University of Tennessee (UT), Knoxville with Ms. Mary F. Poats, Dr. Delcie R. Durham, Dr. Wyn Jennings, and Dr. Larry Goldberg as contract monitors, respectively. Also, the research is sponsored by the Assistant Secretary for Energy Efficiency and Renewable Energy, Office of Transportation Technologies, as part of the High Temperature Materials Laboratory User Program, Oak Ridge National Laboratory, managed by UT-Battelle, LLC, for the U.S. Department of Energy under contract number DE-AC05-96OR22464. The authors would like to thank Mr. D. Fielden of University of Tennessee for machining all samples used in this investigation. Special thanks goes to Dr. R. Belardinelli from Northrop Grumman Inc. for providing Blackglas™ samples.

REFERENCES

[1]R.L. Lehman, S.K. El-Rahaiby, and J.B. Wachtman, Handbook on Continuous Fiber-Reinforced Ceramic Matrix Composites, The Am. Ceram. Soc., p. 495 (1995).

[2]W. Zhao, P.K. Liaw, R. Belardinelli, D.C. Joy, C.R. Brooks, and C.J. McHargue, Metallurgical and Materials Transactions A, 31A, p. 911 (2000).

[3]Nondestructive Testing Handbook, 2nd ed., vol. 10, Nondestructive Testing Overview, Stanley Ness, Charles N. Sherlock, technical editors, Patrick O. Moore, Paul M. McIntire, editors., American Society for Nondestructive Testing, Inc., (1996).

[4]Jeongguk Kim and Peter. K. Liaw, Nondestructive Evaluation of Advanced Ceramics and Ceramic Matrix Composites, JOM-e, vol. 50, no.11, 1998,
(http://www.tms.org/pubs/journals/JOM/9811/Kim/Kim-9811.html).

[5]J. Kim, P.K. Liaw, D.K. Hsu, and D.J. McGuire, Nondestructive Evaluation of Nicalon/SiC Composites by Ultrasonics and X-ray Computed Tomography, Ceramic Engineering and Science Proceedings, vol. 18, no. 4, 1997, pp. 287-296.

[6]H. Wang, L. Jiang, P.K. Liaw, C.R. Brooks, and D.L. Klarstrom, Metallurgical and Materials Transactions A, 31A, p. 1307 (2000).

[7]L. Jiang, H. Wang, P.K. Liaw, C.R. Brooks, and D.L. Klarstrom, Metallurgical and Materials Transactions A, 32A, p. 2279 (2001).

[8]B. Yang, P.K. Liaw, H. Wang, L. Jiang, J.Y. Huang, R.C. Kuo, and J.G. Huang, Materials Science and Engineering, A314, p. 131 (2001).

[9]ASM Handbook: Nondestructive Evaluation and Quality Control, vol. 17, ASM International, 1992, p. 284.

Advances in Ceramic Matrix Composites VIII

KEYWORD AND AUTHOR INDEX